W9-AJK-577

READING WATER

"Drawing upon years of river experience, veteran guide Rebecca Lawton reflects on the timeless lessons that flowing rivers teach. Recounting her journey through roiling rapids—both on and off the river—Lawton's love and knowledge of flowing water shines brightly, leading readers to their own awareness of how river magic can enrich and refresh our lives."

—Tim Palmer, author, *Pacific High: Adventures in Coast Ranges from Baja to Alaska*

Also by Rebecca Lawton

Discover Nature in the Rocks: Things to Know and Things to Do. With Diana Lawton and Susan Panttaja. Illustrations by Irene Guidici Ehret. (Stackpole Books, 1997).

A Capital Discoveries Book, one of a series that features journeys of self-discovery, transformation, inner-awareness, and recovery. Other volumes in the series include:

Ancient Wisdom: A Chinese Calligrapher's Tale by P. Zbar

There's a Porcupine in My Outhouse: Misadventures of a Mountain Man Wannabe by Michael J. Tougias

My Renaissance: A Widow's Healing Pilgrimmage to Tuscany by Rose Marie Curteman

Wise Women Speak® to the Woman Turning Thirty edited by Jean Aziz and Peggy Stout

Midlifeman: A Book for Guys and the Women Who Want to Understand Them by Larry Krotz

Touching Quiet: Reflections in Solitude by Mindy Weisel

Daughters of Absence: Transforming a Legacy of Loss edited by Mindy Weisel

Rivers of a Wounded Heart: Every Man's Journey by Michael Wilbur

Letting Your Heart Sing: A Daily Journal for the Soul by Deborah Tyler Blais

Lessons from the River

Rebecca Lawton

A Capital Discoveries Book

Capital Books, Inc.

Sterling, Virginia

Capital Books, Inc.

P.O. Box 605

Herndon, Virginia 20172-0605

ISBN 1-931868-09-3 (alk.paper)

Library of Congress Cataloging-in-Publication Data

Lawton, Rebecca, 1954-
Reading water: lessons from the river/Rebecca Lawton.
p. cm.
ISBN 1-931868-09-3
1. Lawton, Rebecca, 1954- 2. Women canoeists—United States—Biography. 3. Canoeists—United States—Biography. 4. White-water canoeing. I. Title.

GV782.42.L39 A3 2002
796.1'22'029—dc21
[B] 2002067390

Printed in the United States of America on acid-free paper that meets the American National Standards Institute Z39-48 Standard.

Credits

The following essays have been published before but under different titles or in slightly different form: "In the Shadows of Giants" in *The Gift of Rivers* (Travelers' Tales, 2000); "The Tongue" in *Standing Wave* (1997); "Of Cobbles, Zen, and River Gods" in *Word Tastings* (Santa Barbara Publications, 1998) and *A Woman's Passion for Travel* (Travelers' Tales, 1999); "Faith in the Dry Season" in *Tiny Lights: A Journal of Personal Essay* (Tiny Lights Publications, 2001).

Chapter epigraphs are excerpts from the following publications, each integral to the core of this book:

California Rivers and Streams: The Conflict between Fluvial Process and Land Use. Mount, Jeffrey F. Berkeley: University of California Press. 1995.

Dams and Rivers: The Downstream Effects of Dams. Collier, Michael, Robert H. Webb, and John C. Schmidt. U.S. Geological Survey Circular 1126. Tucson, Arizona. 1996.

Dictionary of Geological Terms, Third Edition. Bates, Robert L., and Julia A. Jackson, editors. Prepared under the direction of the American Geological Institute. Garden City, New York: Anchor Press/Doubleday. 1957.

The Ecology of Running Waters. Hynes, H.B.N. University of Toronto Press. 1972.

Fluvial Processes in Geomorphology. Leopold, Luna, M. Gordon Wolman, and John P. Miller. San Francisco: W.H. Freeman and Company. 1964.

River Channel Bars and Dunes—Theory of Kinematic Waves. Langbein, W.B. and L.B. Leopold. U.S. Geological Survey Professional Paper A22-L. Washington, D.C.: U.S. Printing Office. 1968.

Excerpt from *Cold Mountain* ©1997 Charles Frazier.

Lyrics from "Gonna Take a Lot of River" © MCA Records, Inc., Hollywood, California.

For my mother,
who sent me to the river
in her place

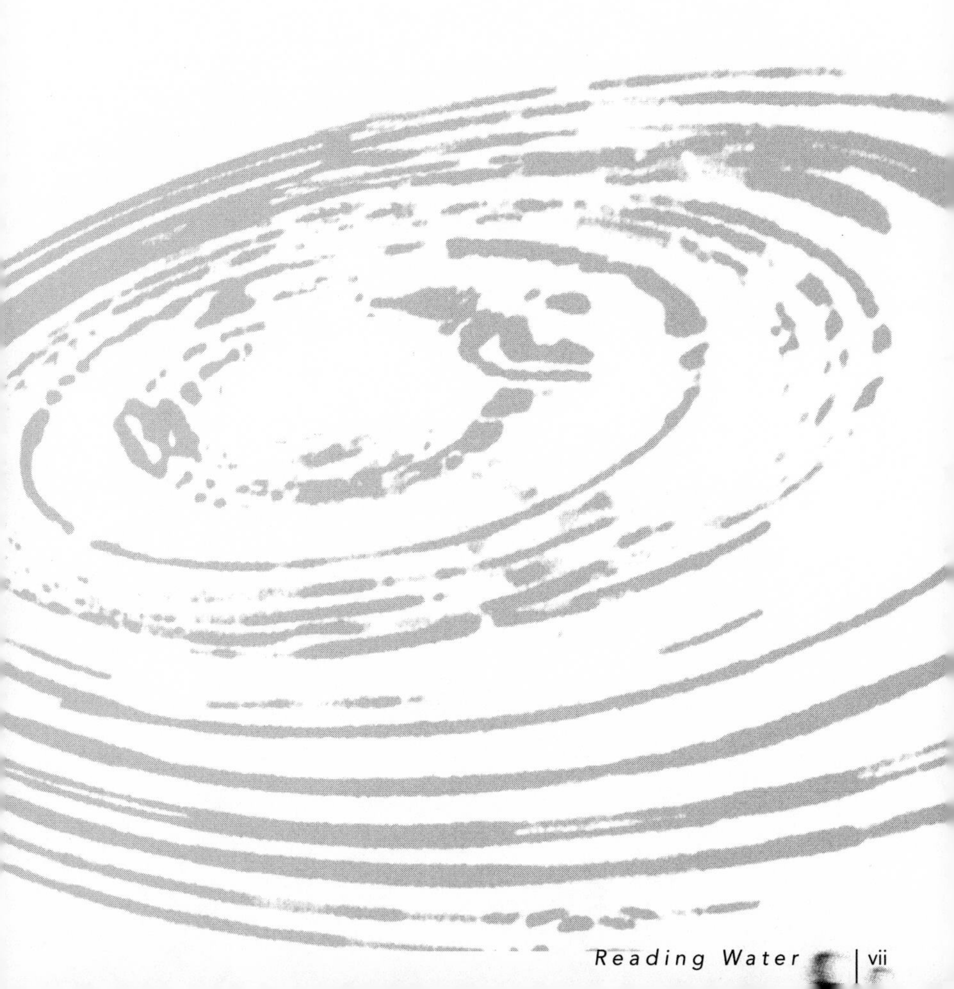

Contents

Preface
xiii

The Thing in Water
1

Braided Love
15

Prodigal River
28

Eddied Out
39

Take the "C" Train
54

In the Shadow of Giants
67

The Tongue
81

Gaining a Loss
97

Reading Water
115

Traveling Bodies
127

Confluence
136

Life on the Floodplain
152

Of Cobbles, Zen, and River Gods
161

Faith in the Dry Season
169

Past the Edge of Land
179

Acknowledgments
197

The study of rivers
is not a matter of rivers,
but of the human heart.

—Tanaka Shozo

Preface

My first view of the river looked like this: a long blue being flexing at the bottom of a steep canyon, where white shallows pulsed in the sun and indigo pools lay back in cliff shadow. Jumbled with purple broadaiea and white fiddleneck, deep carpets of lupine and poppies tumbled down green hills that climbed from the water. The wildflowers shimmered as bright as neon among grasses that turned their blades in the wind. In all my seventeen years, I'd never seen such loveliness. The river was the Stanislaus, a Sierra Nevada foothill stream with a lavish gift for capturing hearts.

That first glimpse of the Stan came to me as it had for many others—as I rode down the Camp Nine Road in the back of a stakeside truck, headed for a river trip. With nineteen other initiates, I rattled over cattleguards and dropped into potholes that jarred my teeth. A rough way to travel, hard on my teenaged dignity. Still, I felt glorious in the spring sun, wearing bikini top and cutoff jeans, making my first trip to a wild river. Courtesy of the truck's driver, who paused automatically at all the right spots, we perched at the final bend on a ridgetop. An ideal spot to gaze at the water. Below us the river frothed in a lively path that we'd soon be boating, the hubbub of whitewater guaranteeing thrills and adventure.

Raised in the city and having just burst out of high school, I ached for wildness, and the lovely, unleashed river instantly called to me. Its fresh water rushed down from high snowfields that in spring melted and poured as runoff through a sweet little marble canyon. The place was drenched in color, not just with its mixed arrays of wildflowers but with streamside canopy of oak, willow, alder, and pine and understory in all shades of green. Intrusions of polished white granite knifed through dark bedrock. Pale beaches lay studded with varicolored cobbles and patches of sand-loving yellow composites and fuchsia Indian paintbrush. Butterflies drifted over the water. A wild beauty.

Mornings filled with the calls and flight of canyon wrens, swallows, mergansers, and red-tailed hawks. Afternoons brought hot upstream breezes carrying the scent of flowering catkins and wet earth. At night, cool winds blew downstream, balm for the soul. As I lay on the sand and watched stars fall, I wondered how I got so lucky and how the world could be so lovely. No master gardener could have planned such a place.

I'd come to the river for a weekend, but I ended up staying for years. Not long after that first trip, I became a professional river runner with an ambition to guide throughout the American West. My wish was granted. In an unparalleled mentorship that lasted for more than a decade, rivers led me to countless unspoiled places, challenged me to be strong, and introduced me to lifelong friends. Moreover, the river taught me to read water—to psyche out where rocks hide in riffles, find safe runs in inscrutable rapids, and keep moving through the flatwater.

Now, years after my life as a guide, I frequently return to the many lessons learned from the river during that time. I may no longer hold a pair of oars in my hands every day, but I'm still steering the craft. Constantly adjusting course. I'm still drawn to wild stream canyons, some of Earth's greatest places. Because, although I've had learned instructors of all ken—in science, literature, art, music, philosophy, love—for which I'm deeply grateful, moving water remains the wisest teacher of all.

—RCL

Sonoma, California

The Thing in Water

We have seen that upstream and downstream migrations are usually initiated by rising water.

—The Ecology of Running Waters

One season early in my guiding career, mad with water lust, I showed up uninvited to spring training in California. Not only had I not been asked along, I'd been specifically discouraged from coming. The first, giddy trips of the season required the services of only a few boatmen, and the training would be strictly for them—or so I'd been told by the Area Manager. The river was still swollen with runoff, a beast gone mad. Its eddies were washed out, familiar marker rocks buried in current. More accidents seemed to happen at high water than at any other time—flips, missed pull-ins, passengers washed from rafts like so much flotsam. Every wave rose to the sky like an unmapped mountain, each trough slid precipitously toward the unknown. Not a scene for novices. Those of us near the bottom of the commercial boating pecking order would have to wait for high water to pass before we got our oars wet.

But a winter's worth of waiting to boat had crazed me. I had to get on the water. I'd finished my first commercial rafting season in the fall, six full months before. Now high water was coaxing me from the shelter of home, beckoning as wind over the sea lures sailors. The early season river had a mighty energy that would be lost by midsummer—I couldn't wait much longer to plug in. To join the unstoppable drive toward the sea that powers every water molecule on the planet. To get on the river, its rapids linked end to end, the breathing room between them gone.

The way to get there, I was sure, was to join spring training that would go on all weekend at the river. Get on that A list of guides. I wanted to stand near shore with the rest of them, the riotous water before us, refreshing my knowledge in rope throwing, map reading, knot tying, and boat rigging. The river, still icy and unforgiving, would wave from a distance like a carnival barker. And we'd be ready for it. If the river level dropped enough that weekend so its current wasn't tangled with logs, we might actually get to go boating.

If I missed that training, though, another season would

pass by like an express bus. There'd be no high-water trips for me, and I might not even get on the guide's B list, if there was one. So I rose in the dark on the day of training and drove to the transit station where the invited guides would meet to carpool to the river. I stood in the cold outside the station, my breath forming small clouds. My heart full of doubts. I could've been at home lying in my warm bed, rather than subjecting myself to possible humiliation and the potential wrath of our boss, the Area Manager.

Minutes dragged by, exaggerated by anxiety as the others arrived and I stood apart from them. Newspaper trucks stopped to disgorge stacks of newspapers at the stands. A few early cars passed, any one of them likely to be carrying the Area Manager. When he finally arrived, driving a fully loaded river truck, he spotted me and did a double take. Our eyes locked. He said nothing, so I piled onto the gear in the truckbed with the others like a bird nesting, settling into my ride to the river and spring runoff.

Boatmen aren't the only beings motivated by freshwater runoff. So are Pacific salmon, who slip to sea on threads of high water from their natal freshwater streams. There are five species of these ocean-going fish: humpback, sockeye, chinook, coho, chum. All names but the first (for the deeply humped back of the mature salmon) come to us through folk etymology from the Salishan language of the Pacific Northwest. Sockeye derives from *séqey̓*, chinook from *c̓inúk*. Coho is an anglicized *k̓ʷə́xʷəθ*. Chum is a slightly evolved *cəm* or *cə́m*, jargon for "spotted" or "variegated," good descriptors for the mature chum.

Evolved names, evolved lives. Each species follows a unique rhythm honed over ages of breeding in its home rivers. Some leave their birth streams soon after hatching; some stay for years feeding on insects and plankton before venturing to sea. Various species spend various times in the ocean, making journeys of several years and hundreds and even thousands of miles. A salmon from California may swim to the Bering Sea in search of a greater food base, then find its way back not only to the watershed of its youth but to the specific tributary where it was born.

Each species returns home in its own time, separate from the others like parallel strands on a loom. Streams hosting more than one type of salmon run on uncannily coordinated calendars, the staggered arrival times providing isolated periods of mating and nesting. Some north Pacific rivers see the chinooks arrive first in June, the chums a bit later that same month. The sockeye follow in early July, and the coho head up in late summer and September. In some watersheds, the humpbacks run home only in even-numbered years, their rotating return one of nature's mysterious phenomena of timing. Then the fish give their lives breeding in their home stream gravels, doomed and sacred torchbearers passing blood and instincts—even the nutrients from their decaying flesh—to another generation of young.

Fresh water flowing to the ocean attracts the salmon back to their streams of origin. Biologists have said that the fish probably navigate celestially while at sea, then are guided by sense of smell into the mouths of streams and up to their home reaches. They have no electronic geographic positioning systems or magnetized north arrows guiding them back toward some lodestone. Something in the water helps bring them home. Could be specific chemicals issuing from the spring-fed heart of the watershed. Could be the sense of some familiar plant or rock type. Could even be some chemical trail left by generations of their ancestors.

This year—right now, in December—my hometown creek is graced by a handful of chinook salmon who found their way into our valley after a few large and early rainstorms. State biologists have told us the chinook are probably hatchery fish who were homing west toward the Sacramento River. The early flows from our creek may have lured them north. The biologists hope these wanderlings can breed here, establishing a brave new generation of salmon.

Years ago, long before my memory of this valley, our creek had chinook. Now, because of reduced habitat, they're seldom seen here, rare and precious as a waking dream. People gather at the bridges to watch the chinook dance their mating waltzes. The fish fan the gravels with their tails in preparation for laying eggs, chase each other through the riffles. Cars stop, crowds collect, children run along the banks for a better look, celebrating the sight of a handful of thirty-inch fish holding in a half-foot of water. At Sunday services, two of the town's pastors announced the presence of the salmon, inspiring members of both congregations to flock to the sidewalk at our Second Street bridge. A friend filmed the in-stream drama, capturing not only the sight of the fish but also the laughter and excited cries of the people.

My neighbors stop me on the street to ask if I've been to the bridge. I have, and they ask me why, of all years, the fish chose now to spawn in our creek. I repeat what the biologists have said, that fresh water lured the fish here.

One neighbor says, "The water did it? Fantastic! I hope they'll stay."

Softly, irrelevantly, I quote Emily Dickinson: "'Hope is the thing with feathers.'"

"Right," he says. "Except in this case, it's the thing in water."

That first spring training in California ran as smoothly as current through a new flume. The river cooperated, dropping just enough to allow for two full runs in which we all shared the boating equally. No drama, no terror, no broken bones. When my turn came at the oars, I rowed as well as anyone, hoping someone would notice and include me in any other early trips. When the roll was called for an elite group of boatmen who would guide high water, I wanted my name to be among them.

My persistence paid off. Soon I'd been added to the early season schedule and was due to start rowing commercially in a few weeks. All my ducks seemed to be lined up. When the time came, and my bags were packed for the first trip, I checked in at area headquarters near Angel's Camp. Strolled into the tiny office. Said hello to the Area Manager and searched the roster to confirm my name.

Devastation. My spot on the first trip had been reassigned to another guide. The early trip bookings were posted for all to see, a litany of the boatman's best and brightest. Which didn't include me.

The Area Manager saw my confusion and disappointment. "The water's going to be really high." His steady gaze was a spotlight on me. "Are you sure you can handle it?"

"I've been on high water."

"Sorry," he said, "but I'm putting on only the old timers."

I studied the roster. Found the name of the guide he'd chosen to replace me. A great guy, but with no more experience than I had. The hell.

It may be unremarkable that my replacement was male—after all, most people rowing trips that spring were. But I knew a hopeless situation when I saw one. I excused myself from the Area Manager and drove back home. Called the main office and asked to be reassigned to Utah. With a new schedule in hand, I repacked my bags—this time for a summer away. Over the next two days, I bummed a ride as far as Winnemucca, Nevada, then hitchhiked farther east. Arriving at a parking lot in Vernal, Utah, I helped rig trips for the Green and Yampa Rivers. Within days I was at the oars as a regular guide on my first out-of-state commercial trip.

No room for me on high water? I'd find some wild rivers to run.

I find it easy to admire the salmon. They live with a faith and generosity that renders them almost holy. They brave the rivers at high flow, travel heroic distances, give their all to return home, and die on the altar of childbirth. Entire early cultures in the Pacific Northwest idolized salmon with great artistic elaboration. Many still do. Craftsmen carve the likenesses of fish-people in wood and ivory, painters immortalize them on skins and masks, metalsmiths work their profiles into prized jewelry. Tlingit, Haida, Tsimshian, Bella Coola, Kwakiutl, Nootka, and Coast Salish people of the Pacific Northwest maintain a worshipful attitude toward migratory fish that is integral to their culture, as central to their lives as drawing breath.

On a trip to the Skeena River in British Columbia years ago, I was surprised to see the dark, round heads of harbor seals breaking the water's slate-gray surface hundreds of miles from the ocean.

"What are they doing this far from the sea?" I asked my friend Doug, a Canadian national.

"Following the salmon," he said. "They're like the Sioux tracking the buffalo. They won't go home until they've gotten their fill of fish."

We trailed the run up the Nass River, north of the Skeena, far back into the bush. Doug had been invited to a totem celebration at a small Tsimshian fishing village. Wild, inhospitable, the Nass valley stretched wide and full of spruce and fir from horizon to horizon. Doug drove us along miles of muddy road, over wooded passes, and along the river, in September already lined with snow and broken by icy boulders. By the time we arrived at the village, we'd long since left the seals behind. "But the people are here for the salmon, too," Doug said. "I'll show you."

We passed rows of modular homes with snow-veneered vehicles parked alongside. Dogs barked at us from the porches, but the place felt empty of humans, with a hush over everything. In the center of the village, Doug and I passed the great cedar pole, raised in public ceremony hours before. It wove the carved faces of Raven, Eagle, Wolf, Hawk, and Bear. As I wondered where all the people were, Doug led me to the door of a huge Quonset hut in the middle of the community. He pulled open the big double doors.

Inside we found a glowing event hall, full of families of Tsimshian. They'd set out more bounty than I could've hoped for at a five-star restaurant, let alone in a lean village many miles from the nearest shipping lane. The room was filled to its curved walls with tables of food. Hundreds of dishes lay out as offerings: herring, roe, cockle, oolican, rice, loaves of fresh-baked bread. Salmon was there, in every imaginable form: shredded and molded in the shape of big fish with olives for eyes; smoked and stripped as jerky; grilled as steaks; baked whole. Doug and I sat family-style at long tables with chieftains and priests, staff from the Bureau of Indian Affairs, officers of the Royal Canadian Mounted Police, the prime minister of British Columbia. Our many conversations filled the building, from the arched ceiling to the sharp corners where the walls met the floor.

I pondered the generosity of these Tsimshian people. Urban dwellers in California pay upwards of eight dollars a pound for fresh Pacific salmon steaks and fillets. We willingly shell out nearly a dollar for a few grams of the nutritional supplement salmon oil. In fine restaurants, salmon is always a good choice—grilled, baked, poached, blackened, or tossed with risotto—but high priced. In this remote village with little obvious sign of material prosperity, the people enjoyed wall-to-wall salmon. They seemed to dig into it with a gusto that belied its presence on their plates every day for months.

On our way home later that evening, Doug remarked on one particularly large salmon that had been served at the feast. The town mayor had dished up morsels of fall-run coho as people passed through the banquet line. "I watched him divvy up that fish," Doug said. "Everyone who asked for a piece got one." As the line had thinned out over about a half-hour of nonstop serving, so finally had the coho. "That salmon really stretched. It was magic—like Jesus and the bread loaves."

Years after leaving California to guide in Utah, I was invited to sit on a panel of boatwomen at a winter river convention in Salt Lake City. I'd recently retired from professional guiding, but during my career I'd been lucky enough to row countless trips in Utah and Idaho and guide ten fabulous seasons in the Grand Canyon. I'd worked such plum rivers as the Yampa in Colorado-Utah and the Selway and Middle Fork of the Salmon in Idaho. I'd had remarkably good fortune, especially considering I was a woman guide who'd started in the early 1970s and had to push through a few barriers along the way.

As I'd gained experience guiding, I'd worked up into first-string crews. The question of whether I was qualified to row high water had faded with time. I'd gotten on plenty of thrilling, early season trips, my days crammed with the excitement and brilliance of free-flowing water, my nights full of the peace I felt sleeping on river beaches. When I boat now on my home streams, I draw on the experience from all those years on rivers throughout the West. Each day of experience was a deposit into a bank of skills from which I now pull interest.

During the panel discussion, a young woman guide asked me if I'd ever been discriminated against during my rafting career. My mind immediately went back to that spring in California when the Area Manager had replaced me during high water.

How to be both truthful and tactful? Over a decade had

passed. "There were a few incidents," I said, "but they were early on."

"What did you do?"

I said, "I kept trying."

"But they won't let me do anything but drive shuttles!" complained the guide. "I want to get on the water."

"Then show up at trainings, camp out at the river put-ins. Work in another state if you have to."

"What if they don't want me?"

"They'll want you. Just don't take 'no' for an answer—and don't give up hope."

Now when the river flows full in its banks, I think of the salmon. Their lives depend on their getting to the runoff, to slide away quickly on the flood. Like unseen thieves, the fish must swim to sea before being caught or killed. High flow becomes their lifeline from spawning beds to downstream pools and ultimately the ocean.

The ones who make it become part of something much bigger than their home watersheds. They travel in worlds Ulysses might have envied. They leave as untried yearlings in sleek bodies, to return fully equipped for adult life, with deeply humped backs and elongated, predatory jaws. The sharp-toothed mature salmon appear more like barracuda than the innocent, trout-like stream fishes that departed years before. They've changed their dress as well, casting off the silver slips of their youth for green and scarlet spawning livery, with spots and stripes for homecoming.

Similarly, new boatmen and women must get to the river to dip their oars in spring runoff. The smell of the mud entrained in water, the breaking foam in a washed-out hole, even the rain stealing down drainages to the river—all are the siren's call to the thrill of high water. Those who make it are ready to hold the power of the water cycle in their hands. Their return is triumphant, their bodies the dreams feasts are made on.

Braided Love

The steepest place is

where the channel divides around an island . . .

—Fluvial Processes in Geomorphology

When I was seventeen, I fell for the assistant guide on my first river trip. We met and shook hands while he was scouting Death Rock, the first major rapids on the Stanislaus. He was McGav, the trip's blond, blue-eyed baggage boatman. I'd come as a guest, my head already spinning from the river's wild beauty and the dozens of introductions to new people that morning. As we shoved off from shore, and by turns disappeared into the rock garden of Death, I snuck glimpses of McGav from my seat on a paddle raft: the glamour of him, a real river guide. The dip and swing of his oars as he ran sweep. The relief to my teen-aged soul when he emerged past the final monster boulder and it was clear he could row.

The river was at its best then, in the late spring when wildflowers still blanketed the hillsides and lined the white sand beaches. The Stanislaus ran cold, with clear snowmelt barely warm enough to be called liquid. McGav and I skipped stones from the beach at Chinese Camp in the evening. We discovered the canyon wren's song from across the river, the cliffs with maidenhair fern dripping spring water. We divulged how we both hungered for wild things and adventure, how the river fed both. In each other we saw the longing possessed only by the young in such full measure.

After the two-day trip, our paths diverged. I went to college, studied minerals and dinosaur bones. McGav dove wholly into river life. He moved to a small silver trailer on the banks of the Stanislaus—our river—near a deep green eddy upstream from Parrott's Ferry Bridge. In all seasons, all weather, he rafted and kayaked. When he wasn't boating, he hiked throughout the watershed, wrote me letters about what he'd found: the skins of animals, remnants of old placer mines, grinding stones of the long-gone Miwoks. He told me of full days, as well as evenings with wood-fire dinners in the clearing outside his trailer. Often he had visitors, drawn to his good spirit, great campfire cooking, and the constant promise of a tent sauna followed by a naked plunge into night-cold water.

By the next spring, when I was training to become a guide myself, McGav's reputation had preceded him all over the river. He was the mountain man who'd wintered at Parrott's. The fearless one, who'd inner-tubed the Stanislaus at high water. Who'd boated past all the commercial camps while standing naked on the bow of a friend's raft. A lunatic.

Right then, at the height of his river fever, our romance began. It started on another Stanislaus trip—this time, he was a full-fledged guide and I rowed baggage. The trip ran on an overcast spring weekend, with everyone in wool caps, wetsuits or long johns, and—whether or not we were on the water—huge Mae West style lifejackets over the top of everything. At lunchtime the first day, we pulled into the mouth of Rose Creek, a miracle of a bedrock sidecreek with sweet water pouring into plunge pools. Alder catkins hung in plump, sweet tassels, spreading musky scent. Lovers' incense. Even in the cool air under the cloud cover, the fragrance of new growth filled the air.

Everyone eager to explore, we hastily made lunch and scattered like windblown leaves up the sidecanyon. Munching a sandwich, I walked a ways and chose a pool to sit beside. The musical flow of the water lulled me into a hypnotic state. Glued me to the spot. After some time of solo revelry, I noticed McGav ambling up the creek. He hadn't yet seen me—instead, he seemed fully engaged in studying each pool as he wandered. Occasionally he'd toss a stick into the creek to see which current it caught.

Unlike the rest of us, still in our Mae Wests, McGav had cast off his lifejacket and stripped his wetsuit to his waist. He seemed immune to the cold in a polar-bear way. He paused beside a riffle, tossed a stub of willow to navigate the tiny waves, and stood stock-still like the statue of a god—broad in the shoulders, slim in the hips, confident in his strength. My heart drummed, a thrill locked in my chest.

McGav looked up, caught me watching, grinned his impossibly wide grin. His blond hair, always unruly and in his eyes, spilled down over his forehead. He brushed aside a few locks like a bead curtain. Tossed away the last stick he'd picked up for testing creek current. Then, with a sort of wry expression growing on his face, he closed the distance between us, stood before me a minute, and bent down for a kiss.

In my mind the passion that began on Rose Creek and the Stanislaus is still linked with the river, the river, the river. Huddling together on streamside beaches beneath McGav's warm blankets purchased in real Mexico, protected from morning dew or night frost. Swimming like otters in the sun-warmed river pool down by his trailer near Parrott's Ferry. Learning to Eskimo-roll a kayak with McGav standing by in an icy eddy, refusing to give up until I got it.

The planet continued unnoticed in its orbit. The moon circled, time spun by. Summer bloomed sudden and hot, the river guiding season full upon us. McGav took a job rowing the Grand Canyon. I transferred out to Utah canyon country. With him on the Colorado River and me far upstream on the Green, we could only trade letters and telephone calls. Twice or three times we were both off the river the same day and arranged for dizzy desert meetings in Arizona or Utah. We packed our time full, always on the move hiking, driving, dancing. Each encounter was short as cloudburst, always ending with me standing somewhere on a sidewalk or in a parking lot. Never sure when we'd meet again. Waving good-bye as he drove away.

When fall came, I returned to college, but rivers still dominated my thoughts. I studied fluvial geomorphology, scoured numberless references on stream mechanics. Lost myself in the wisdom of river gurus, discovering the two principal types of channel patterns—*meandering* and *braided*—that define the shape of the river as seen from above. God's view.

The meandering channel, with a single thread of flow, loops and winds in great, mathematically definable serpentine curves. The braided channel has many streams of current separated by multiple bars or islands. Dividing around islands and rejoining to divide again like the cords in a plait of hair, braided streams run in patterns that are neither predictable nor repeatable.

Braided streams lay down sediment of all sizes. Finer-grained silt or mud covers coarse-grained sand; cobbles mingle with boulders. When viewed in cross section, as in roadcuts, the sediments of ancient braided streams show many vertical changes—sandstone, mudstone, pebbly conglomerate, then sandstone again. The meandering stream, on the other hand, deposits sediments that grade steadily upward from coarse to fine—boulders at the bottom, then cobbles, then sand and silt. The meandering and braided river models filled my many waking hours, their differences haunting me as a lover ghosts through dreams.

McGav joined me on weekends but found my preoccupation with school dull. My college mates and I were too bookish in our humor. "Why read about rivers when you can be on them?" he asked. His eyes took on a distant look. Soon he told me he was through visiting for a while.

"I'm sorry," he said, firing up his grandmother's steel-blue Chevrolet Apache.

"When will I see you again?" I asked.

"I'll call you." He headed down my street, disappearing over the hill that sloped toward the ocean. A flock of gulls laughed overhead. Beyond the curve of earth that he'd driven down, winter swells rose up in the sunshine. My heart fell away after him, its sudden sinking a terrible, unknown feeling.

Later, I spilled out my sorrow to a friend. In the midst of consoling me, she had a revelation.

"Hey," she said. "What sign was he, anyway?"

"Sign? Gemini," I replied.

"Ah, the Twins," she said. "No wonder."

To *braid* comes from the Middle English *breyden*, meaning to move suddenly. The Middle English term derives from the Old English *bregdan*, to draw (as a sword). Both roots imply abrupt movement. An animal leaping, startled by sound. A duelist unsheathing his blade in a flash. The concept of sudden movement applies well to the braided stream, which will abandon any of its channels with eye-popping swiftness. Usually the change comes about because the braided stream is a high-energy entity, capable of carrying great loads of sediment. When dropped as the stream flow wanes, the sediment clogs one or more channels. The main component of current pushes abruptly into a new bed. Channel switching.

Many of my favorite boating rivers have braided reaches that plunge headlong out of the mountains—the Selway and Salmon in Idaho, the Green and parts of the Yampa in Colorado, California's gold country rivers and coastal streams. Braided and meandering reaches alternate in concerto-like fashion—fast, slow, fast—although the meandering reaches outnumber the braided by about ten to one. The river braids as it descends steep slopes, meandering later through the soft alluvium of a middle-elevation valley and lower coastal plain. On the many meandering stretches, we ride lazily and catch our breath on anticipated patterns of flow. On the breathtaking, rare braided runs, we plunge on mad rides around islands, plummeting in surprise descents over submerged boulder fans. Neither predictable nor boring, that ride on a braided stream.

Recently I fell for another Gemini man. This time I believed I could handle him, because there wasn't a river in sight. There was no chance we'd be diverging and converging all over the West. Plus, I'd learned so much in the twenty-five years since I'd lost McGav that I believed my judgment to be improved. I'd gone through college and graduate school, changed jobs and careers more than once, had the experience of a few serious love affairs, married, had a daughter, divorced, recovered. No chance I'd choose someone likely to hurt me, no matter what his astrological sign.

Still, I felt wary. "I've been hurt by a Gemini before," I admitted to my friend Susan, a wise woman schooled in astrology.

"Well, I probably shouldn't say this. But Geminis lie."

I thought about it. "No Gemini's actually *lied* to me."

"I don't mean like fibbing. But you know—they're two beings in a single body. Always torn two ways. They can't stick to one story."

"So should I avoid him?"

"No," she said. "It'll be good for you. You can check out your Gemini karma."

Soon I dove into a love affair with my Gemini friend. He was funny, smart, and nice looking, and he seemed too grounded to float off in some heady direction. Raised in a small college town back east, he'd kept friends from high school, college, and graduate school and had maintained a solid career and community life for twenty years. Our adventures together took place fully on land—picnics, ski trips, jogs through the woods.

Susan's words didn't come back to me until the romance ended. Seems my lover had been with someone else all along. He just couldn't choose between us. In shock, I broke things off and escaped to the Sierra, traveling up Highway 80 on a Saturday before first light. Near the snow line at Gold Run, I remembered to check out the highway roadcuts as I zoomed by. Steep slopes filled with ancient river deposits, stunningly beautiful and right next to the road. Surprising, especially on drives like that one, when I reached them just at dawn. Appearing red because of the weathered iron in them, the rocks glowed even rosier in the pink sunrise.

Gawking at the Highway 80 gravels, I recalled that they're stream gravels known to be laced with fine gold—the same sedimentary rock that the hydraulic miners of the mid-century Gold Rush exploited with high-pressure hoses. Huge meandering rivers are said to have deposited the gravels at a time when the mountains were not undergoing much uplift. Back then, there were no new, steep slopes for the streams to charge down.

Perhaps the rocks are the leavings of ancient meandering streams, and I bow to the greater knowledge of those who've studied them in detail. But at dawn on that Saturday, with me licking my love wounds, I could've sworn I was seeing evidence of braided streams. They're mostly cobbles—the same coarse sediments laid down in high-energy states—with no obvious fining-upward, meandering-stream sequences. The roadcut shows plenty of vertical variation, too, evidence that the stream changed its course frequently. Channel switching.

"Liars," I muttered, as I continued up the mountain.

I should tell you that McGav did call again, true to his word. The startling fact was that when he resurfaced he was ready to settle down. He hadn't decided with whom yet, but he did me the honor of seeing me first. Five months of utter silence, then his call came at four a.m., the ringing of the telephone gradually worming its way into my sleep. I wrapped myself in a blanket and stumbled to the kitchen in the dark.

"Hello, lover," said the voice I'd known so well. McGav said he'd been talking with another boatman in a bar. "I told him how I'd left you, and he accused me of cutting and running."

"You did."

"I couldn't decide what to do. We were headed different ways."

He said he'd moved to the mountains, up to a boatman's property near the headwaters of the Stanislaus. He'd sold his trailer and bought a tipi. He'd been living alone there except for the company of a borrowed malamute. Together they'd skied and backpacked all over the Sierra, two characters in one epic story.

He came by on the weekend, perhaps believing he was visiting the same girl he'd left months earlier. I wasn't her, of course, although I didn't know it until he stood on my doorstep. He grinned hugely, looking great in a plaid summer shirt and straw cowboy hat. We went for a drive in his Apache, sitting at far ends of the bench-style front seat, following the same arch of road toward the beach he'd driven months before.

How different it would have felt to ride with him back then. How my heart would have swelled with the joy of being by his side, watching the entire sun-filled ocean come into view. But sitting up front with him now felt hollow and faithless. with no way to rejoin after suffering such a wide split. Parked by the beach, we neither moved closer nor said much, and later that day we parted for the final time.

Within weeks he met his future wife hitch-hiking near a ski area in the mountains. They hooked up immediately and married soon thereafter. When I saw them together, they seemed a good match, and they've stayed married through whatever changes must have come.

Had I been wiser back when McGav returned, I might have understood better how love goes, how it tolerates rifts. Because if love were a streambed, it would have to be braided some of the time. Made of many channels, it's the stuff of steep rivers. No matter that braided geometries are outnumbered by the more staid meander ten to one. Maybe love descends madly for a while, casting off burdens like cobbles and sand as it travels forward, laying down groundwork like islands and later settling into a nice, single bed. Maybe then there's no need for the precipitous rides and channel switching.

Maybe.

Prodigal River

The river emerging from a dam is not the same river that entered its reservoir.

—Dams and Rivers: The Downstream Effects of Dams

A sprig of dried plant lies on my kitchen table. Sent to me by my sister, Jen, the specimen has five sky-blue blossoms pressed flat and dried in a fold of paper towel. Leaves the color of green tea branch like monkey arms from the flower's pewter-gray, four-inch-long stems. Desiccated, mahogany-brown roots mass in a jumble at the base of the stems. The specimen is a wild-flower, *Nemophila menziesii*, or baby blue eyes. Jen collected it more than twenty years ago in the Sierran foothills for her college Botany 103 class. In a letter she reminds me that I was with her the day she picked and pressed the flower: "We decided to take a walk up the north bank of the Stanislaus River. Remember? The water was low and the rocks exposed out in the river. But we weren't looking for rapids that day, we wanted wildflowers."

Despite the age of the parched specimen, it's clearly baby blue eyes. Petals of pale sapphire fade at the base, joined together mid flower in a central white eye. It's a lyrical plant with a lovely, cup-shaped corolla. When alive, the flower turned generously upward. Caught dew, offered itself to insects, bared its delicate, shivery stamens to breezes. Jen's notes from our walk say that the hills above the Stanislaus were covered with the blossoms. "The flowers favor 40-degree slopes in sandy loam with surrounding live oak and gray pine forest." Among hundreds of such places, we chose one, and my sister set out a 20- by 50-foot test plot.

What had appeared at first glance to be a homogeneous patch of hillside among "classic yellow-pine forest" in fact proved quite diverse. In the shade of oaks, we found ethereal surprises. Fairy lantern and brightly colored fiesta flower. Ferns and saxifrage in moist soil at the base of granite boulders. Out on the sunny slopes, lupine and grasses grew among popcorn flower and fiddleneck. Jen noted "a mottled mosaic of yellow and white."

Her test plot startled us both in its variety and wealth, like hidden treasure. Jen collected individuals of twenty different species of wildflowers from inside the plot and thirty more from nearby. She archived the fifty specimens as "a sample flora from the Stanislaus River near Parrott's Ferry Bridge." Recently, during a move, Jen came across the specimens and mailed the baby blue eyes to me with a letter: "I thought this might make you smile. Because of that hike so long ago, baby blue eyes always remind me of you."

During my first trip on the Stan in 1972, while I sat with a group of guides and passengers in a cool, huge limestone cave near Chinese Camp, I learned of New Melones Dam. The latest U. S. Army Corps of Engineers' dambuilding project, New Melones, had by that time been in preconstruction for six years, many miles downstream at the historic Old Melones townsite. The dam was to be 625 vertical feet of rock with an earthen core, a design like a jelly doughnut—dense material with a soft center. If filled to the brim, New Melones would back up 26 miles behind the dam, creating California's fourth largest reservoir, with a surface area of 20 square miles and 100 miles of fluctuating shoreline.

I received this news in utter blackness in the Chinese Camp cave. "But what will happen to this canyon?" I asked.

Whoever replied had a broken voice. "Flooded. Completely gone."

No way, I thought. No one would deliberately destroy Eden. California seemed to have more than enough dams, and even as a teenager I had a sense of what we'd lose. There were just no other rivers as lovely and accessible as the Stan. By the early 1970s, its upstream whitewater run had become the most popular in the state. The project seemed as inconceivable as building the Pyramids in the middle of Main Street.

But time went by and work progressed on the dam, deep in a downstream canyon. Men with hardhats scraped hillsides clean with bulldozer blades. They diverted the river through tunnels in bedrock. Dug footings with backhoes. My colleagues and I began to accept the building of the dam as inevitable but hoped for a compromise with the Corps: a filling plan that allowed for water storage only as far upstream as the Parrott's Ferry bridge. This "build-it-but-don't-fill-it" strategy seemed acceptable to us—better than ruining the entire canyon. And it would meet the Corps' professed need for water, power, and flood control. The reservoir wouldn't be half empty, it would be half full.

The river hung in the balance as debate fumed over New Melones. A group known as Friends of the River materialized in a grassroots groundswell and led an effort to limit filling the reservoir. As a guide on the Stan, I took part. Each trip through the canyon assumed not only a bittersweet feeling but also a political charge. In our ammo boxes, we guides carried pencils and petitions to the state legislature, collecting signatures from California-registered voters at the take-out for river trips. On our days off from boating, we sat with our petitions at card tables in front of grocery stores in Sonora and Columbia. Or we traveled to San Francisco to stand on city streets and beg signatures from passersby. We must have looked a bit backwoods, young men and women in blue jeans and T-shirts. But we were effective. We needed 300,000 signatures to get on the ballot. We collected 500,000.

Next came public meetings, a push to get out the vote, and a whirlwind publicity campaign in support of our initiative, Proposition 17. Friends of the River drew on $238,000 in contributed funds to run a media campaign that consisted largely of radio spots. We were opposed, however, by a powerful, deep-pocketed coalition. They spared no expense. Saw our 200 grand and doubled it. Funded such propaganda as a mammoth billboard at the San Francisco approach to the Bay Bridge that read, incredibly, SAVE THE RIVER, VOTE NO ON 17.

How shocking it was to see such deliberate humbug plastered across the landscape. Voting YES on 17 would actually save the living river. A NO vote favored New Melones, meaning the end of life in the river canyon as surely as if Noah's flood were to wash through. It floored me. Our opponents were lying—the evidence of it clear to my teen-aged eyes.

Baby blue eyes. They not only grew like weeds on the slopes of the Stanislaus, but they graced the banks of some of the other rivers Jen and I came to know. We found them on our visits to the American, Merced, Tuolumne, and Yuba. We found them on southern Oregon rivers, the Rogue and Illinois. Our time spent outdoors led us to full-time guiding, with my sister and me both training and hiring out as boatmen for commercial trips. We became like human otter, spending all day every day on or near the water. Rivers held our full attention and devotion for years.

A diminutive woman determined to handle big boats, Jen was able to guide using a foot brace to raise herself on the rowing seats. It worked beautifully, as if she'd been born to it. She rafted high and wild water on canyon rivers in Utah and held her own on crews of rowdy boatmen, even leading trips for many years. Eventually we both guided in the Grand Canyon, well into our thirties, long after our family and friends had given up hoping we'd live productive lives. We were goners, tuned to rivers and seasons. Summer, with its hot, full trips scheduled back to back. Fall, a time of poignant separations and endings. Winter, with its cold, long months waiting for the rivers to flow again. Then each spring, ready to cast off our landlocked lives, we converged on the river. Often we began with an inaugural trip down the Stan, when it flowed strong and cold and the hills were draped in wildflowers, a voiceless constituency waiting to be counted.

Unfortunately, reservoirs with all their benefits come at a great price. A dam changes everything about a river, flooding essential habitat for animals, inundating prehistoric rock paintings, submerging historical townsites. Trees drown where they stand. Sediment collects in oversized reservoirs, starving the downstream river of necessary surges of sand and silt. Fish populations cannot tolerate the drops in water temperature that occur in reservoirs, with disastrous results.

Californians can speak from hard experience. Here, dams are held responsible for decimating half of our native fish species and fully 95 percent of our wild salmon. With unstoppable drive, we've leashed water to the tune of 1,400 dams throughout the state. For decades, the undammed river was considered profligate, untamed. A wasted stream, with falls begging to be harnessed for hydroelectricity, channels waiting to be fitted with locks for ships. Any dammed river a working river.

Turns out, the wild river is the champion of working rivers. It feeds a myriad of lives, from insects to ouzels to beaver, all part of an important creature community. It waters the flood-plain, replaces depleted soil. It builds the very land we inhabit, depositing deltas inch by hard-won inch at the edges of continents. It flushes marshes and slices sloughs. Flowing into the ocean, the river creates a freshwater wedge that pushes into saltwater, nourishing the life-giving water chemistry at the end of the stream. It carves out the deep, green world of estuaries, provides the safety of harbors. The river's glorious, generously given water flows to the ocean, replenishing the very fluid that feeds our delicately balanced system of earthly life. It's as important to maintaining equilibrium on earth as the mountains that rise from the mantle.

Nothing in a free-flowing river ever goes to waste.

Those of us who battled against New Melones will never forget it. Until the defeat of Proposition 17, I'd never really considered that liars could win. My mother always said that cheaters never prosper, and I'd believed her. So what had happened? *We* were the guys in the white hats. *We* were telling the truth about the place. In spite of our honest effort, or perhaps because of it, we lost on the night of the election. Unbelievably, the returns came in against the Stan, so fast that those of us watching didn't even have to stay up to watch the outcome on television.

After the election, the environmental battle raged on. Over more than a decade, river lovers tried everything. Appealed to President Nixon to halt construction. Brought a lawsuit against the Corps for its flimsy Environmental Impact Report. Finally petitioned for a re-vote on Proposition 17. The ultimate act became the instances of civil disobedience at the eleventh hour, as waters rose behind the dam. Defying the laws that had cleared the reservoir footprint for inundation, a few true believers handcuffed themselves to rocks at the river's edge. They spent critical days and nights chained like criminals in the canyon as their messengers disseminated word of their actions. The nation held its breath as the filling halted, then breathed a sigh of relief when the defiant individuals were found and removed from the canyon.

Filling resumed. Homesteads slipped under the inky, oceanic quiet of the waters. The Old Melones townsite went under, as did canyon walls with hematite-paint pictographs dating back eight thousand years. Riverside limestone caves submerged, with their underground sanctuaries big enough for a crowd. The showcase stalactites and stalagmites we'd taken pains not to ruin with our touch began to dissolve instantly under millions of gallons of water. Protesters formed a human chain across the lower river to mark an acceptable upstream limit for the floodwaters. The water rose above it. Soon it became too high for anyone to safely stand in, and all we could do was bear witness.

Death by drowning is never a pretty sight, and the Stanislaus River canyon was no exception. Swallows wheeled in panicked flight at the Old Melones Bridge as water engulfed their mud homes, still full of nestlings. Small mammals, stranded by the rising flood, clung to the tops of willows and had to be rescued by boat. Water, normally a wondrous, life-giving substance, oozed into every dry cove, onto each warm patch of sand, into the nooks in every living tree trunk. Jen's plot eventually flooded. Whole forests drowned with it.

Last week, remembering the Stan, I stopped in Santa Barbara, California, to visit a tree that survived the deluge. A gray pine, the tree was transplanted as a seedling from the Parrott's Ferry area of the Stan just before the filling of New Melones. One of the few pieces of riparian life to escape the flooding, it now stands at a roadside park a mile or so from Highway 101. Since its replanting around 1980, the pine has grown to about 30 feet tall.

One of several trees in an island park surrounded by suburban roads, the pine lives in a posh neighborhood. Its boughs reach toward the mountains, its needles catch the morning sun. No doubt its bark sometimes glows red at sunset, as before. It may be the only gray pine from the Stan that's still living and breathing. Like us, it's a refugee from the wild river.

I'd rather my daughter—and all the children—could see the Stan's grass-covered hills and the wildflowers and gray pines in their natural homes. I still reminisce about that rich place, back when my sister and I could barely count the numbers of plants and animals that lived there. Jen's notes from our collection day say that "someday, when the Stanislaus River is nothing more than a silt-filled reservoir, someone will ask, 'What was it like before the lake?'" And we'll know.

We have the gray pine—and the baby blue eyes—to always remind us.

Eddied Out

Water within an eddy tends to have a velocity well below that of the rest of the river.

—Dams and Rivers: Primer on the Downstream Effects of Dams

While training to be a Grand Canyon boatman in the 1970s, I managed to work a few trips with my older brother, Tim, already an accomplished river guide. He'd been rowing the Colorado professionally for years, his quick wit, wiry competence, and obvious passion for the natural sciences earning him favor among not only his passengers but also his fellow guides. Hooked on the outdoors all his life, Tim had first encountered river running while on a backpacking trip in the Canyon. As soon as he'd glimpsed the rubber rafts shooting rapids where the trail emerged at the mouth of Horn Creek, he was a goner. "That's for me," my high-school-aged brother had said, nodding to the boats that rode the waves of the Colorado like pleasure craft on some rowdy conveyor belt. He couldn't get the sight out of his mind. Upon returning home, he'd summoned his courage to write to a commercial river trip outfit, asking to be trained as a guide. A few weeks later, he got a call back that said, "You wrote us a good letter. How'd you like to come to work running rivers?" By the next summer, my brother was rowing boats for a living.

Throughout his commercial boating career, Tim invariably found ways to get the most from what the river had to offer. For one, he loved a wild ride and liked to handle his boat accordingly. Not that he took chances—he just knew what he and his raft could handle and wasn't afraid to push the limits. Every run, he'd split the twin holes in Sockdolager, meaning he'd run down a thin line of current racing between the rapids' two midriver monster reversals. Finding the line between them is tricky, like negotiating the furniture in a dark room, but with much higher stakes—a little to the left, you hit a hydraulic bigger than the black hole of Calcutta. A little to the right, you find a reversal the size of Moby Dick. Most boaters avoid the middle of Sock altogether by rowing far to the left. Not Tim—he'd go for the juice every time. He'd aim his boat for the narrow, smooth slot between the holes, hoping to climb the watery crest surging between them. Invariably, he'd hit it dead on and emerge out the downstream side exultant as a kid out of school.

Tim also preferred to run the Canyon's first big rapids, House Rock, straight down the center. He'd plunge through the upper waves and continue through the middle toward the maelstrom of the bottom hole. The rest of us began our approaches to House Rock with a ferry maneuver that moved us to the right throughout the rapids, but Tim would hang it out there farther left. Speeding through the wave train, gathering momentum like a car with no brakes, he'd stay midriver until the very last minute. When the bottom hole loomed mere yards away, reaching with watery hands to grab his raft, he'd pull right with all he had. Spray would sizzle up from the tossed depths, suction would try to claim him, and his boat would spin—barely missing the colossal maw that threatened to drag him into the drink with all his trusting riders.

As an assistant, I was rarely able to ride with my brother, from sheer lack of space in his raft. Each boatman rowed an even six riders, the raft's limit, and each day Tim quickly filled his quota. I jumped on wherever there was room every morning, rounding out the last odd lot of paying passengers, but there was no need for that in his boat. The *bona fide* riders (affectionately called "the people"), wearing sun shirts and carrying cameras and knapsacks, would line up for a seat in his raft. My brother seemed not to notice them vying to be with him—he was busy being at the height of his form as a guide, giving the people their money's worth. They sucked up his knowledge of canyon geology and wildlife with awed expressions, and they looked transformed by his rides through the huge rapids. Executives and managers would start the trip with faces sealed with the knowledge of their own importance, but after a few days with my brother in the Canyon, they looked as open and shining as children.

I was an impatient assistant. By the time I'd arrived at the Canyon, I believed I knew what I was doing—I'd already guided professionally in California, Utah, and Idaho—and I figured I knew more than enough to row the Colorado. "I can do this," I told myself, as I rode through Badger, my first real Canyon rapids. The waves of Badger, small by Canyon standards, were still bigger by far than any I'd seen on other rivers. "I want to do this," I thought, as we continued through Soap Creek, and I saw that bigger didn't necessarily mean trickier, just stronger. By the time we reached House Rock on the second day, I felt confident. "I know I can do this."

But still I had to get in my training trips. So I rowed—the boatmen resting on the baggage like pharaohs, ready to jump for the oars any time we reached significant water. And I rowed—keeping a sharp lookout for eddies, which in the roiling water of the gorges seemed to wait at every bend to trap a wayward raft. And I rowed some more—arms burning—as I wondered how long until my true time came.

Eddies are a mixed bag. By definition they're currents of air or water running contrary to a main stream. Often the currents travel in circles, whirlpools of the elements. Originally the eddy was named *ydy* by Scotsmen and *itha* by Old Norsemen, inhabitants themselves of a wet, windswept world. Beyond the broad labels given eddies, however, generalizations about them stop. No two are alike. They may be quiet, shallow places in rivers where water shunts from the mainstream to lap in whispers against a sandy shore. Or they may be as turbulent and tossed as a windy ocean—deep, swirling cauldrons that threaten to suck boats in and never spit them out. Some eddies are huge, sluggish countercurrents laden so thick with trapped driftwood that they dare us to walk across them, floating bridges to shore. We needn't be fooled—the flotsam won't hold. Beneath it lurks a watery trap into which we can fall.

Even in a mother lode of metaphor as rich as the river, the eddy stands out as a precious diamond. We can mine its real-life analogues endlessly, using language rooted in being stuck: *backwater* towns, *dead-end* love affairs, *blind* alleys. Most sticking places are as easy to get into as the river's eddy; most are similarly difficult to leave. On the river, they're everywhere. Eddies form at the tops of rapids, at midpoints, and at the bottom, and there's no end of ways they can trap us. We may be aiming for them, our hearts and muscles fully engaged in trying to bag them. We may be ignoring them, focused as eagles on something downstream. We may be trying to avoid eddies with every fiber of our beings, on or off the river, and still we find them.

Created by obstacles to the main streamflow, eddies swirl behind midstream boulders, trees fallen into the river, spits of sand and cobble reaching from shore. Water fills the space behind these barriers as matter rushes toward a black hole. An eddy may consist of a quick sidecurrent, a whirlpool circling behind a rock, or a narrow, near-shore backwater hiding from the greyhound-swift mainstream. An eddy's speed depends on the velocity of the current swirling in and around it, its shape determined by the size of the barrier behind which it hides, like a hunkered-down trout.

Between the eddy's reverse flow and the river's main current lies the eddy fence. As diverse in nature as eddies themselves, fences run the gamut from placid to violent. On a tame eddy fence, the water simply and quietly flows downstream on one side and upstream on the other. In a powerful fence, vortices surge many feet high between two ripping countercurrents. Creating a sort of purgatory for flow, water on the fence swirls up, down, and in circles but never really goes anywhere. Sooner or later, though, the current has to commit in some direction and get off the fence, spinning in toward the eddy or out toward the mainstream. But while on the fence, the current remains in that room between heaven and hell. When a raft strays near or directly onto an eddy fence, one of two things can happen: the boat will slip downstream on a prayer, or it'll be sucked out of the main current, onto the fence. Then it's into the eddy for a go-round or two. Or ten. And it's up to the unfortunate boatman to get back out.

On the Colorado, eddies reign supreme. They're fierce, enormous, greedy—they could suck the Queen Elizabeth off course. Currents on their eddy fences boil so high, you need a stepladder to see over them. Some of the biggest even have titles. King Edward. Ever Eddy. Helicopter Eddy. I never could keep track of which name went to which killer eddy, and indeed there seemed to be plenty of debate about it, but boatmen innately agreed that place names applied just as well—the eddy at Bedrock, the eddy below Deubendorff, the last eddy before Lava, and so on. Boatmen seemed to agree on something else as well: once you've bagged an eddy—or it has bagged you—you're said to be eddied out.

Granite Falls in the Grand Canyon is a narrow, thrilling ride, with waves running so daringly close to a bedrock wall on river right that the run might have been dreamed up for a theme park. Except that the rock is harder than plaster, and the chances of actually hitting it good. It's a terrifying sight to see bedrock a mere ten feet away as you scream through the waves at breakneck speed. The falls carries boats past the wall most of the time, but nasty sidecurrents wash right into cold, hard rock. You can begin Granite at the right of center and run near the wall all the way, or you can start a bit left. Either way, by the end of the wave train, you'll be somewhere near the wall. Worse, if you happen to pass it safely, you still must contend with Granite Eddy, which lurks on the bottom right like the sirens laying for sailors.

Granite Eddy has a precipitous fence that will spin your head on its spine both coming and going. Boatmen try to stay off the fence, but it's not easy. If you enter the rapids near the wall, chances are that your downstream momentum will carry you past the fence, but you might hit the wall, perhaps breaking your right oar in the process. If you choose instead to enter the rapids a bit left, you might not collide with anything, but your angle is more likely to slide you toward the fence below. There are no guarantees. But that's the key to Granite, knowing that the price of cheating an eddy with a fence reaching to the moon may be a quick glance off bedrock at thirty miles an hour.

My brother may have been the guide who clued me in to the secret of Granite Eddy. Or another informative guide may have told me about it, or I may have figured it out for myself. Even if Tim didn't pass me that particular piece of information, he may as well have—I learned so much about boating from him, the line between what I picked up from him and from elsewhere blurs. He took me through my first whitewater, Cadillac Charlie on the Stanislaus River, when I immediately decided boating was my thing. He showed me how to spin a boat gracefully in flatwater, demonstrating it himself, with easy and synchronous tugs at the oars. He was the first boater I saw run the left wall at Crystal Rapids—years before high water changed it enough to make the left run comon. He showed me where vireos nest at the mouth of Stone Creek, where to find the Great Unconformity in layers of Canyon rock, and how to leap daredevil-fashion between canyon walls in upper Saddle Canyon. It follows that he may also have told me how to deal with Granite Eddy.

On one of those early river trips in the Canyon with my brother, I was riding with my colleague Bangle on the afternoon we reached Granite. We rowed the sweep boat, the raft that brings up the rear, and Tim ran one boat ahead. All day Bangle and I had gotten glimpses of Tim's runs. He'd gone for the right horn at Horn Creek, snuck left at Crystal, hit everything straight in Hermit—all beautifully—and we knew he'd run the wall at Granite as well. We watched as he set up on the far right at Granite, in that pregnant moment before the first drop. We watched as he waited for gravity to claim him, hovering, his people holding on and poised expectantly. When his raft plummeted into the waves, sinking into the first trough, rising up on the next crest, Tim swung the oars as the raft rose and fell. He hit the main wave, the second, the third, then came off the fourth a little off-angle. That was the moment of doom. With a sideways jolt, his boat slid out of the mainstream and fell like a rock into Granite Eddy.

"Uh-oh, he's in it!" Bangle yelled. The people's heads swiveled, their faces wildly puzzled. Bangle was already set up for a left run, and before he could explain anything, we plunged into the rapids. We slipped down the left-side waves, feeling the pull of current dragging us right. Still, the wall was a respectable hundred feet away, so we held our breaths. We alternately climbed waves and fell, climbed waves and fell, until we passed the threat of the wall and rose to the top of the biggest roller. We cheered with elation, as if we'd topped some remote peak.

Then, at the crest of a smaller wave, our luck changed as fast as an alpine storm. We stalled, spun, and dropped down the wave's backside, landing at a perfect angle to slide into the eddy. "Son of a—!" Bangle screamed, reacting quickly to pull left. He stroked once, twice, many times, as we bumped along the fence toward its downstream terminus where the current made its final split.

We rowed snout boats in the Canyon back then, two-ton, 22-foot neoprene rafts that we retired guides blame in retrospect for our current back troubles. The dinosaurs of oar boats, snouts are made of Army-surplus bridge pontoons, one end of each tube curved up like a rodent's nose. The design of the snout resembles that of the modern cataraft, with a deck tying two parallel tubes together Huck-Finn style. With relatively little floor to drag on the water, the snout is fairly nimble for its weight. It's not a bad invention—big enough to carry lots of supplies, constructed of readily affordable surplus gear, and tolerably rowable. But loaded with six people and twelve days of gear, the snout is as heavy as a freighter.

"It may be a pig, but it sure can track," claim the snout's

proponents, who were more numerous in the 1960s and 1970s when alternative craft were scarce. To track means to follow a narrow course, as a sprinter keeps to a single ring in a race, and the snout does it well—once headed in a given direction, it generally continues that way, and its forward momentum can be diverted but not stopped. Add to the snout's natural tracking tendencies the fact that in rapids passengers like to ride its J-shaped noses for the wet, rough thrill of their lives, and you've got a tracking machine that won't quit. With two 100-plus-pound people riding the bow, the front-end weight drags the rest of the snout around like a little red wagon.

On the Granite Eddy fence, moving the oars as fast as he could, Bangle worked against the tracking snout. Even without riders on the noses, the eddy sucked us toward it, capitalizing on our angle and claiming the sheer weight of our load. In slow motion, without effort and at the last minute, the current sorted us right.

We had no choice but to ride the upstream flow of the eddy to the top, where Tim's boat still swirled in the backcurrent. As we floated closer, his people called us over, asking for help. I jumped onto their boat and discovered that no one was at the oars. Instead, I found Tim lying crumpled and exhausted against the front splashboard, the people surrounding him with concerned expressions and open canteens. I quickly learned that he wasn't hurt, just spent. Rag-doll limp, his arms rested at his side. His face looked uncharacteristically humbled. Granite Eddy had recirculated his boat four or five times. Each time he'd tried unsuccessfully to cross the powerful eddy fence. With his full boatload of riders, he'd given it everything he had, but he couldn't get out of Granite Eddy.

Time spent in an eddy can feel like time wasted, a maddening delay to forward progress. Eddied out, you've lost precious momentum. Time spent in an eddy can also be seen as respite, a place for our souls to catch up with our bodies. On a winter river, eddies are quiet enough for first ice to settle. In spring, the shelter of eddies lures fish to linger and ducks to flock. Kayakers dancing through rapids use the slower current of eddies to help them pivot, pause, and decipher the water ahead. Around them, the river moves downstream like a gale-force wind, the backcurrent a haven. Eddy out, and you buy yourself time to breathe.

As it happened, Tim's stay in Granite Eddy burst me out of my backwater of assisting. Maybe an emergency should not be seen as opportunity, but that's when my opening came. I must have been ready for a break, like the Buddha preparing for *nirvana*, because I only balked for an instant. Then I climbed onto the center seat and took over rowing Tim's boat.

I didn't succeed in getting out of Granite Eddy on the first try. Circling more than once, I called up all my strength and cunning for each attempt to break out. Most likely Tim coached me that, with an eddy fence so fierce and the main current rising above the countercurrent like a barricade, I needed to build momentum to get out. Probably I took his advice, set up a ferry angle that would slice me upstream into the current, and charged the top of the eddy with a youthful will. No doubt I failed on the first few attempts, only to be slammed back into the eddy to try again. I know, because I've been there many times since that first encounter, and that's what happens. When I finally broke out of that watery cell, it must have been because my sweat and determination happened to conspire at the right time with the whirlpools on the eddy fence to let me out.

Tim stayed as popular as ever on that trip and on others to follow. He enthusiastically resumed his thrilling runs the next day, after a good night's sleep on the beach at Boucher Rapids. He didn't guide professionally much longer in the Canyon, because his talent as a teacher of natural sciences was pulling him toward graduate school at Stanford University. He's now mid career as a university professor and brilliant researcher of geology.

As I was beginning my stopover as a Colorado River boatman, he was ending his.

Now I know that my days spent as an assistant Grand Canyon guide and apprentice to my brother were the true time I'd been waiting for, a great stage of life. Why I was in a hurry to get my own boat and become a regular Canyon guide baffles me now, but no doubt it had plenty to do with my being twenty-two years old, eager, and a female wanting to test limits to boot. But I'm fortunate to have waited and trained as I did, because in all my rushing from river to river for experience, those days spent in limbo allowed me to reclaim my abandoned soul. Time filled with the honor of tenured boatmen sharing their love of the river, bend by bend. They patiently explained the rapids, taught me the names of rocks and grottoes, passed on history and lore. None of us knew it then, but they were handing me a kingdom.

I do remember the result of breaking out of Granite Eddy. We continued downstream to Hermit Rapids, the biggest, smoothest, most fun waves on the river. Tim rode up front with the others, urging me on. Exhilarated by our escape from Granite, I entered Hermit with a profound sense of release. Showered by the cold spray, cheered by my brother and the people, I knew that something had ended and something just begun.

Take the "C" Train

Communities change, either wholly or in part,
as the current, measured at one time or water level,
varies from place to place.

—The Ecology of Running Waters

Despite my best efforts to stay out of Teepee Hole, we're being sucked toward it like logs to a timber chute. My raft with its sole passenger is lined up perfectly for the brown, crashing current that's fast approaching and less than one hundred yards downstream. We're going to have to run the frothing mess, a reversal of flow that collapses back on itself in the middle of the rapid's main wave train. Accepting our fate, I'm pushing on the oars to move us into Teepee, building momentum. In fact, a spectator happening on the scene at this instant might guess that I'd intended to run the hole all along. Forget that it's a brisk morning when no one wants to get wet, half an hour before the sun will light the water and warm the canyon. Forget that a monster boulder lurks just beneath the hole. And never mind that for all the awake I feel, I should still be wrapped in my sleeping bag back in camp. We're committed.

From experience, I know that Teepee is a keeper, a watery trap to be avoided anytime. The river dances in it like droplets in a hot pan, shaken and imprisoned. Boats that enter the hole share the fate of the water, becoming one with it. How will my passenger react when we drop into it? Will his head spin and eyes pop? Will he register anger, joy, surprise? He has the bulk of two men—can his weight carry us through if we land dead center in it? What if we flip and he has to swim?

We charge over the brink of the hole and bump down with a jolt. For a moment we continue to track and it seems we'll continue downriver with no problem. Praying we won't be stopped, I keep working the oars to move us forward, dipping the blades quickly like the paddles of a windmill. A brief time passes in which I seem to be effective, until the boat gives a telltale shudder and abruptly halts. For one awful moment we hover, then we're drawn backward toward the heart of the hole. There's where the nasty business can occur—we may get sucked down, engage in endless spins, or overturn in a sudden, breathless flip.

My passenger remains relaxed and facing downstream, as motionless as a mannequin. He seems to be unaware of our upstream creep, or he may consider it just part of the show. Or he too is still waking up, or he's too petrified to move. I figure he'll be alarmed, though, when I climb over the gear and walk past him to the nose of the boat. But no—when I do, he doesn't react, as he doesn't react visibly when I grab the front D-ring and step off the bow into the water. Quickly I lower myself into the cauldron of churning current. Submerged except for my head and forearms, I hold tight to the D-ring with both hands. No way I'm letting go of it. My body and legs catch a tendril of current that's streaming downriver, and the boat with passenger intact follows me out of the hole.

I climb back into the raft, wet but feeling heroic, finally awake.

That evening in camp, my fellow guides congratulate me for the neat trick in Teepee. One guide especially is intrigued—Michael, an inquisitive person with an intense Van Gogh gaze. "Throwing out a human sea-anchor," he says. "Where'd you learn to do that?"

I could reply, "At spring training, where the senior guides told me, 'When you're stuck in a hole, take the "C" train.'" Because they did tell me that.

"Meaning?" I'd asked my trainers.

"The current! Get something—anything—into that downstream current. It'll pull you out."

My answer to Michael has to be more complex than that—it wasn't enough for someone to simply instruct me. Rather, I have to give credit to a litany of killer holes that have claimed me or others before me: "Skull, Crystal, Phil's Folly, Clavey, Warm Springs, Lava Falls, Satan's Gut, Widowmaker—"

"Right," Michael says. "Well, those places must've been in your nightmares. You looked asleep when you jumped off the front."

"Maybe I was asleep." My move to the bow may have been instinctive, an unplanned response contained within my cells, more destiny than decision. How else to explain a move smooth as a dream, slow as a waltz?

Practice, practice, practice. Al, my jazz ensemble teacher at the University of Utah, ended every improvisation session with that advice. I took in his words, my face burning at how poorly I played the guitar in my lap. Determined to do as he said, I still floundered each time my turn came to solo. His words reminded me that all the talent and teaching in the world come to nothing if not cultivated. Practice and intentional experience are key.

Every night I took home my instrument and played the standards Al had set before us—"So What," "Autumn Leaves," "Satin Doll." Fingers to strings, I deconstructed chords and strung their notes together into chains of rudimentary improvisational melodies that I prayed would grow more sophisticated with time. Faithfully I mapped out each solo in my head and practiced, practiced, practiced. Starting slowly, I'd build a line of music, then set the metronome faster and faster to see how much speed I could handle. Often the flow of the metronome's ticking would sweep by me, leaving me behind.

During jogging breaks in the neighborhoods at the base of the Wasatch Mountains, I tried again to find a rhythm I could catch. Humming the chord changes to "Green Dolphin Street" and "All the Things You Are," I pounded the sidewalks and streets of Salt Lake City to forty-year-old melodic progressions. Today the music is still linked in my mind to the mountains' snowy couloirs flushed with alpenglow, brick houses with white shutters and in-law basements, barren terraces rimming the mountains.

But in the harsh fluorescent light of the practice room the next day, my little melodies would crumble under the unyielding beat of the ensemble's momentum. Al would stroll among us and listen without emotion as we struggled with our instruments. A suave man in polo shirt and slacks, he stood with arms folded over his chest, his concentration a spotlight on each soloist in turn. As I hammered out mush, day after day, his expression never changed. I gave him credit for not cringing, at least not within my sight. And I pledged myself to even greater amounts of practice time, as my face still burned.

At the end of a day of boating on the Yampa River, I sit near the water, contemplating reversals. Otherwise known as holes, reversals are places in the river—like Teepee Hole—where the current stalls on the downstream side of a barely submerged boulder or rock ledge. The water hovers a while in the hole, crashing back and upstream. In the clear mountain streams where I learned to boat, the boulders gleamed underwater like submerged faces—the shining white of granite, the dusky red of chert. On the dun-brown Yampa, though, which brims with silt and fine sand, the holes show up only as mounds at the river's surface. They look a lot like the smooth, rolling shoulders of water marking the tops of innocent wave trains.

A raven crosses from our camp to the far shore, and my eye is drawn to the opposite bank. Shadow chases sunlight up the buff-colored sandstone cliff while the pink glow of sunset grows in the canyon. After the direct light leaves the rock walls, colors still shine off the river's surface in a lustrous rainbow. Then the many colors on the water give way to reflected salmon pink, then crimson. Soon night settles over the river, and the midstream holes become only dim humps. I strain to study them anyway, until I believe I could find them in the dark.

Practice, practice, practice. Al continued to end every ensemble rehearsal with those words. My fingertips developed thick, impenetrable pads of callous that allowed longer and longer practice time at home. In the evenings I jogged through the neighborhoods to the rhythms and melodies flowing through my mind and heart. At night I played my guitar until I fell asleep with it in my arms. Then every morning under fire in the ensemble room, I'd painfully trip through an improvised solo, unable to get into the stream of music rushing past me.

Practice, practice, practice. I'm sure the words apply to other artistic endeavors as well. To meet the muse, we must show up for it on blind faith, as Romeo showed up in Juliet's garden. But unlike Romeo, we must be there time after time to see what will happen. We have to be ready to be swept away, or the courtship is doomed to fail. Why shouldn't it? Even if we stumble upon the great treasure we seek, without experience and readiness we may not even recognize it, and it will pass us by.

My favorite rivers are full of holes that have stopped me or my friends abruptly in our paths—Clavey Hole on the

Tuolumne, Crystal Hole and Satan's Gut on the Colorado, Warm Springs Hole on the Yampa. We've watched each other go awash in these holes—recirculating in Clavey, flipping in Crystal, dropping backwards into the Gut. We've skirted Warm Springs Hole with no room to spare, in fact dangling our bows over the gaping maw of churning water while praying we wouldn't get sucked in. Lava Falls in the Grand Canyon finishes with some monumental tail waves that sometimes collapse back upstream as rough reversals, and my colleagues and I have run those, too, countless times. Usually a raft running the falls can't miss Lava's tail waves—a boater just has to punch through them as straight as possible and hang on for the ride. Sometimes they break on you; sometimes they don't.

One time I hit the tail waves just as they decided to crash, and they stopped my six-passenger, two-ton snout boat midstream. We surfed as if we weighed nothing. The river screamed on downstream all around us, eager to reach the bottom of its thirty-seven-foot drop, while the boat skimmed in place, made weightless as a beach ball. Snatching the oars from my hands, the current pinned them to the sides of the boat. There was nothing to do but ride it out, and we hung for many moments in the reversal, water mounding and towering past the nose of the boat. The sun backlit the water purling above us. A seven-person excursion on the Banzai Pipeline of the Desert Southwest, it was also first-class river surfing.

Just when we seemed to be lingering in the wave-hole forever, we twisted a notch sideways. Someone yelled, "Hang on! We're going over!" and the boat tipped up on a back tube. Rubber hovered over us, blacking out the view we'd had of wave and sky.

Wondering why I'd want to hang on if we were going over, I fell off the rowing seat into the boiling water below us. I had to swim the rest of the tail waves, Himalayas of water though they were. Only after my passengers had pulled me back into the boat farther downstream did the events of the incident become clear. It seemed that I was the only one the river had wanted. Having exacted its human sacrifice, it righted the raft and finally let it go into the downstream current.

Skull Rapids in Westwater Canyon, Colorado, has a hole to be reckoned with, too, and I had the honor of plunging into it my first run through. A boatman's plan entering Skull usually involves avoiding the hole, which comes up fast near the bottom of the rapids. Despite my following the traditional strategy of starting midstream and cranking left, I crabbed an oar at the entrance and tracked directly into the hole's open mouth. My passengers and I dropped quickly into it like children tripping at a dead run. I was mortified—an army of boatmen watched from shore, and I knew I'd be the talk of the town that night when we all got off the river. *If* we got off the river.

We stayed in that bad reversal for entire minutes, not the usual three seconds exaggerated by disaster, but two or three full minutes, as the raft twirled, filled with water, spun, and pitched like a demon in the hole. My passengers screamed and staggered around the lurching boat, characters stuck in a B-movie nightmare. Advising them all to jump, I resisted my own powerful urge to abandon ship. But no one did. Instead we all sloshed around interminably, holding dearly to something—a D-ring, a line, each other—until one of the passengers finally fell into the river. He grabbed onto the downstream side of the raft, pulling on it as the river tugged him downstream, and the raft popped up and out of the reversal like a beast set free.

After Skull and Lava, I vowed never to be stuck in a reversal again.

Weeks at the University of Utah turned to months. Months grew to semesters. Classes ended, summer passed on the river, and the school year began again. I showed up again for ensemble despite my discouragement, not sure I could stay committed much longer. This Romeo was looking like a no-show.

Once again leading our jazz ensemble, Al continued to watch without expression as he strolled among us. Sometimes he applauded those for whom things seemed to be working; the rest of us he listened to carefully but tight-lipped. And he always sent us home with the advice, "Practice, practice, practice."

Then one morning in the middle of "Green Dolphin Street," Al spoke to me over the music. "Try this," he said, as my solo approached. "Play just one note per beat for sixteen bars. Keep it really simple, really sharp."

I closed my eyes. The music that had grown so familiar flowed on. I waited for my cue, then plunged in, playing as Al had told me, one note per beat. I didn't so much sound the notes as place them on the stream moving past me, a quick four-beat of musical waves carrying the pieces of the song's chords. It wasn't easy, letting my fingers hammer out this unheard, unplanned tune, but I forced myself to stay with it. Allowing the rapid key changes to pull the melody from within me felt like a plunge down a very steep slope. But I'd already leapt, so I fell, finding somewhere in my soul the lost advice that said *get something—anything—into that downstream current.*

Sixteen bars passed in a moment that not only lasted forever but also ended in an instant. I opened my eyes. The ensemble had frozen in place, instruments still to fingers and lips, but with music silenced. Coming out of my musical sleep-walk, I was surprised to see Al standing before me, a huge grin on his face.

"Congratulations, kid," he said. "You pulled it off."

Finally awake, I looked around for confirmation. It was true. The other musicians set down their instruments and applauded.

A few weeks following my early morning encounter with Teepee Hole, I'm back at the boathouse, writing out food orders for an upcoming trip. Michael of the Van Gogh eyes is just returning from a trip on the Yampa. He pulls a trailerload of muddy river gear into our boatyard and parks near the ware-house, where he can offload equipment. Before he moves a single item, though, he crosses the yard to say hello. I'm resting on an old bus seat under a tamarisk verandah in front of our boatman's trailer. Many miles distant, thunderheads shift above the bare rock of mountains, and the light changes dramatically.

Michael stands before me, the shadows dancing far behind him. He focuses his blue-eyed stare on me. "I landed in Teepee Hole this time," he says, then quickly adds, "actually it was intentional."

"Why do a thing like that?" I ask.

"To try the human sea-anchor trick."

"What? You're nuts."

"Maybe. But it worked like a charm." He appraises me as if he's a rookie pilot reporting in to Chuck Yeager.

I'm not sure how to react. After all, I'm a pretty conservative boater, not one to drop into reversals for thrills. "Wouldn't it be better to just stay out of the darn thing altogether?"

"What?" Michael looks scornful.

"Row around it."

"I could've, I suppose." In a rare break in intensity, a grin crosses his face like a light. "But it was fun," he adds. "And I needed the practice." Then he turns to unload the gear from one more river trip.

In the Shadow of Giants

A river or drainage basin might best be considered to have a heritage rather than an origin.

—Fluvial Processes in Geomorphology

"You couldn't run this in a million years."

Andy, Debbie, Miles, and I stood overlooking Great Falls, a formidable set of cascades on the Missouri River in Montana. On our way to canoe a stretch of the river, we were stunned at the sight of the falls. It was no Niagara, but still it fully spanned the river, with no navigable sneak chutes.

"I thought Lewis and Clark paddled this river in wooden boats," I said. "How'd they get past this?"

Meriwether Lewis and William Clark had traveled *up* the Missouri, which we'd heard was not much for rapids. But we weren't in search of whitewater anyway. As commercial guides, we'd been running big rapids all summer for many years. All we hoped for was to extend our typical four-month boating season into autumn. We knew little of the Missouri, except that a stretch of it winds through the Breaks, a high-plains badlands near the Canadian border. But we had maps and the name of a canoe outfit in Fort Benton that could set us up to paddle downstream more than 150 miles to the Fort Peck reservoir. These were enough. We'd made Montana our destination.

Quickly, we'd closed out the season's books and warehoused our rafts, oars, river boxes, and bags. Cottonwoods had shivered amber in cool drafts off the Uinta Mountains as we'd stuffed our river bags into Miles's dark-green Pinto station wagon. Knowing we had to move fast before true winter hit, we'd driven straight through for a day and a night, stopping to sleep for only a few hours near a high pass in the Yellowstone. On the second day, we'd reached the town of Three Forks and detoured to the Missouri Headwaters State Park. Although I'd expected to see your classic headwaters, a tiny spring issuing from the earth amid mossy boulders and maidenhair fern, instead we'd found a quiet confluence in an open, sunny plain with grasses long since turned brown by summer heat. There the Jefferson, Gallatin, and Madison Rivers flow together to form the Missouri.

The interpretive signs read that Lewis and Clark, with their band of discoverers, arrived at Three Forks in July 1805. They'd already traveled up the Missouri nearly three thousand miles from St. Louis, wintering over along the way in present-day North Dakota. When they reached the Three Forks confluence, they recognized it as the river's headwaters: all forks measured nearly 90 yards wide, so to call any one the Missouri would have been to give it "a preference wich its size dose not warrant as it is not larger than the other." After naming the streams and scrambling for a few days to figure out which one to pursue farther west, the party continued up the Jefferson toward the Continental Divide.

Piling back into the Pinto, we'd proceeded on to Great Falls. We'd scoffed at reports that called the falls unrunnable. Lewis and Clark had portaged it, but that was expected in an upstream run. Plus they'd been burdened with heavy boats and supplies—nearly two centuries before, no less. In the modern age when boatmen could navigate just about anything in rafts made of high-test materials, we wondered how great Great Falls could be.

Imagine our surprise to see rock ledges rising like giant stair-steps in the riverbed, cascading with frothing, muddy water.

Each shelf of rock posed a barrier over which you could easily crash and burn. Even worse, though, was the prospect of a portage. Great Falls dove through a bedrock canyon bordered by boulder fields and sloping fans of talus that rendered prospects of a carry formidable. There was simply no easy way around it. To run the falls, you'd have to be a fool or a hero. To portage, you'd have to be a god.

"Jeez," said Miles. "Maybe Lewis and Clark were badder-ass guys than we thought."

We stood in silence, humbled in the shadow of giants. We remembered what we had forgotten: that immense spirits precede us—always—when we travel on rivers.

That Missouri trip was the beginning of my affinity for Meriwether Lewis, and it has stuck through the years. Perhaps not because he somehow dealt successfully with Great Falls, or even because he was handsome, accomplished, and competent, which history says he was. Instead I'm drawn to rumors of his mental illness. Not many of our national heroes have taken their own lives, but Lewis did. He was apparently prone to depression that in the end overwhelmed him. So I empathize greatly—I too have fought the demon depression, although it's a beast that seems smaller than Lewis's.

His biographers have said that Lewis pretty clearly was manic-depressive, meaning that he dealt with surges of euphoria and invincibility alternating with waves of nausea and utter helplessness. We have it on Thomas Jefferson's own good authority than Lewis fell into "sensible depressions of the mind" that Jefferson believed had genetic roots: "Knowing their constitutional source, I estimated their course by what I had seen in the family." While Lewis made his westward journey in 1805, though, he seemed fine—in fact, neither Clark nor the enlisted men documented odd behavior on the part of Lewis. Lewis did neglect his own very important journal writing from September 1805 to January 1806, speculatively because of depression, but he never failed the expedition. Instead he conducted himself admirably for the entire eight-thousand-mile journey by river, on horseback, and on foot. In his musings after Lewis's death, Jefferson wrote that the strenuous daily activity and responsibility of leading the expedition must have put off Lewis's usually troubled mind.

I'm no expert on manic-depressive disorders, but I know a bit about depression in general. In short it's caused by brain chemistry out of balance. Low levels of the compounds serotonin, dopamine, and norepinephrine bring on feelings of worthlessness, low energy, and fatigue. We'll probably never know what Lewis's specific brain chemistry was, but I often wonder if it could have been managed by a modern course of drug therapy, psychotherapy, and diet monitoring. Proper treatment, had it been available, might have eased his agony.

There's no telling, but there is truth to tell on what depression feels like. When I meditate on Lewis, his suffering feels familiar. I'm with him as he reflects in a funk on his failures at age 31, writing that he had "as yet done but little, very little indeed" in his life (although he had just successfully reached the Continental Divide, discovered the Missouri headwaters, met peacefully with the Shoshones, and traded for all the horses he needed to continue west). I shiver with empathy as he writes desperately of his failed attempts to marry. I agonize over his struggle the night he took his life—his pacing restlessness as he prayed that Clark would come, his shouting arguments with himself, his emotional end. His final hour, which must have felt so private, has become shared posthumously, a book creased open for all to read. For me and for thousands of others.

News of our Missouri expedition preceded us down the river. During the day, farmers and ranchers wandered to the shore to say hello. In the evenings they politely left us alone. River residents treated us like honored guests, offering fresh drinking water and vegetables pulled up from late-summer gardens.

It's big country, and we were moving fast. We discovered that we could rig sails by stringing ponchos between paddle shafts held vertically in each canoe. Then we were carried like torn clouds over water. Because the river flows east in a country of prevailing west winds, we sailed fifteen, twenty, and twenty-five miles a day downstream on the same breezes that would have blown us upstream on our home rivers across the Divide. We weren't in a hurry, but with the going so easy we couldn't help sailing as far as the winds would take us.

Such stark landscape. As Andy put it, "It looks like we're on the moon." Not so much the White Cliffs area, with its walls and spires that resemble those of the rock canyons of our home rivers, but the downstream part of the run. The Missouri Breaks. There, the slopes of soft alluvial badland are interrupted only by little runs of juniper stringing up gullies. These verdant jags cut away from the river like dark-green scars that lie mostly hidden to the boater looking downstream. Here and there a cottonwood crowds the river, the only large trees for miles. As we sailed through the Breaks, we watched mile after mile of sere grassland curve away from us along the arc of the earth.

At night, after we'd settled in camp and stopped moving, black moods gripped me. I grappled with an internal dialogue I'd grown so used to during bouts of depression. What in the world will I do this winter? Who will I be once I'm off the river? Some evenings the melancholy came on right after dinner, and I slipped into silence.

Debbie asked me about it. I replied, "I'm feeling an end-of-season gloom."

"Yeah," she said. "The back-to-the-real-world blues."

"You too?"

"I've had it before. Always in September."

By morning each day, though, any storms in my mind had broken. They seemed to clear as we packed up the canoes. Gone with the night, they stayed away all day, never plaguing me as I walked by the river, or rigged our poncho sails, or cleaned up camp in the crisp twilight before bedtime.

In the months leading to his suicide, Lewis abused himself badly with drink and narcotics and couldn't focus on his responsibilities. His state of mind seemed to be made worse by his depraved lifestyle and inactivity. Money slipped through his grip without stopping, and he couldn't hold the interest of the young women he courted. More than one hoped-for marital engagement ended in a potential bride terminating her connection to Lewis. Appointed to a government position, he wasted the resources of his office and failed to carry out his daily duties. Over all hung the specter of his meticulous expedition journals, which, to Jefferson's dismay, Lewis couldn't seem to polish and publish.

Had he been engaged in the "constant exertion" that Jefferson knew calmed him, could Lewis have staved off his "hypochondria"? Could he have functioned well again, married happily, written and published his account of the westward journey to riches and acclaim? Jefferson blames the "sedentary occupations" set before Lewis after the expedition for the return of Lewis's troubles. I don't dispute Jefferson's diagnosis, but I believe that Lewis not only needed to be challenged, he needed the river itself. Boating on a gentle stream has been shown to boost serotonin levels in the brain. Rafting whitewater supposedly elevates dopamine and norepinephrine. Whether moving up the Missouri or down, or wandering on foot in search of new discoveries, Lewis was never far from moving water. The river was his drug, administered in ways that may have kept him in chemical balance for as long as it lasted.

None of us who traveled on the Missouri likes to sleep in tents, so each night we lay out in sleeping bags under the stars. River dew invaded camp as a silent intruder, waiting until the deep hours to steal over us. Normally we rested snug in our bags, immune to any chill, but being restless and often visited by nagging spirits after midnight, I sometimes awakened during the dewy pre-dawn hours and felt the cold.

My strategy at such times is to sit up and meditate. I pull my waterproof river bag close for a backrest and draw up my sleeping bag. The practice settles my mind, transforming sad feelings to a blank weariness that eases me back to sleep. As I sat in the dark awaiting sunrise on the Missouri, my mind relaxing, I felt the calm that hung in the air. The coyotes had long since stopped howling for the night—in fact, the air sounded so still I could have believed they'd never wailed at all. Andy, Debbie, and Miles slept on, unmoving lumps visible under a waxing moon. The lunar set was still a short time off, and I decided to walk the river. I dressed in several layers while still in my bag and slipped out into the night.

Once up, I followed the riverbank. Moonlight brightened the ground well enough to illuminate my way along the water, which shone with a muted glow. The incised gullies of the Breaks rose and fell across the river, casting shadows. I moved upstream, following a sweet, familiar scent I'd picked up on the wind. The breeze smelled of ripening rose hips. I stopped, transfixed by this symbol of the high country where wild roses grow with other untamed things. If the rose hips really were there, ripening on uplands, perhaps a group of explorers laboring slowly upstream after months on the hot plains in 1805 also had caught their scent. If so, along with the river growing clearer nearer to its source, the rose hips would have conveyed great promise and the nearness of mountains. Perhaps, like the river, the appeal of maturing fruit gave Lewis and his men reason to keep going.

The breeze blew colder and I shivered. I decided to run back to camp. The moon was sinking fast, and utter darkness would leave me alone and lost in unknown country.

Packing up the canoes in the morning helped my mental state, as did focusing on general camp activities such as cooking and washing. While doing chores, my internal chatter ceased. My thoughts turned to the Missouri, and my mind filled with the simple tasks to be accomplished to proceed down the river. The days teemed with names and places: Lundy Ranch, Virgelle Ferry, Coal Banks Landing, Marias River, Judith River. Clark named the Judith for his cousin Julia Hancock, whom he later married. Lewis named the Marias for his cousin Maria Wood and the "pure celestial virtues and amiable qualifications of that lovely fair one." Perhaps he wished to wed his cousin, too, but Lewis was not to marry in his short life.

While he was on the Missouri, Lewis did not fail. Although he didn't find an all-water route to the Pacific, it was only because it did not exist. If such a route had been anywhere within ten thousand miles, Lewis would have found it. He did explore the river's headwaters, of course, the great source of the river. Not only the confluence of young streams at Three Forks, but the place in the mountains where the waters originate. His journey absorbed him completely, keeping him functionally healthy for two and a half years. It brought him together with a corps of discoverers who trusted him thoroughly, and it joined his name for all time with Clark's. In the great-hearted middle of his journey, Lewis pledged himself "to live for *mankind*, as I have heretofore lived for *myself*." At the time he had no idea the weight of his words, how they'd live through centuries. He couldn't have appreciated just how many lives he would affect, and indeed he has touched millions. Anyone who knows rivers and the American West knows Lewis. Through his strenuous explorations, even in his aloneness, he found a point of origin, the place where we inner-are.

For all the wild country we were accustomed to traveling as river guides, the Missouri Breaks counted among some of the most untamed. The northern prairies stretched in all directions, rolling up to vast Canada less than a hundred miles north. The coyotes sang in packs that sounded big and diverse. Beavers worked at night, tail-slapping the water as we drifted off to sleep. Over everything arched a bowl of stars wider than the world. Sometimes I lay awake pondering the same sky Lewis measured with his sextant in 1805.

Orion stood among the constellations, one obvious sign that we'd shifted seasons and moved under a winter sky. In my fits of wakefulness, I contemplated reaching the end of our time on the Missouri and returning to the real world— a deep, black pit of a feeling. On the trip's last night, I tossed and turned, sat up to meditate, and focused on the darkness inside my eyelids. Time passed, and I felt calmer. Soon I zipped out of my bag and rose to find drinking water.

A light shone in camp. Andy and Debbie had cleared the kitchen table. They sat on ammo boxes on either side of it, playing cards in the glow of Andy's candle lantern. I gravitated toward the light like a lured insect.

"Join the club," said Andy. "No doubt Miles will be arriving soon, too."

Barely twenty yards away, Miles rolled over and propped up on an elbow. "What're you playing?" he asked.

"Hearts," said Debbie. "The cure for restlessness."

"Deal me in," said Miles.

"Me, too," I said, pulling up my ammo box.

We played for hours, not talking about the coming winter. We'd left the river before, we could do it again. How hard could it be, anyway, to return to civilization?

The Tongue

The tongue coincides with a sharp increase in slope, which translates directly into an increase in velocity . . .

—California Rivers and Streams

It's time I told the truth about that day on the Rogue River. I've embroidered the tale, changed names and details, and made myself the hero—most likely in hopes of remedying the many mistakes I made. It's true that after the accident, all I wanted to do was stay home and chop wood. As if I could cut away what I'd done. After all my other chores, I'd go outside and hack and saw until dark, bit by bit splitting all the oak rounds in the shed. As I worked I kept seeing the river, how wild it was the day we ran it, rising hard in the eddies. Trees sliding by, big and torn up, rocks and mud still in their roots. The whole flight of steps from the boat dock folded like a ladder toward the lodge, pulled up and tucked away.

One night after about a week of all my woodchopping, when I was tired but still swinging hard, pushing, sending chips flying, I missed the stump and just grazed my left boot. That scared the piss out of me, so I had to quit for a while. I sat outside watching the wind shake boughs in my big cedar tree. Wondered how much longer I'd have to spend at the woodpile before I'd ever want to get on the river again.

Driving to the boat ramp with Michelle the morning of our river run is still fresh in my mind. She was the girlfriend of a friend, and she wanted a lift down the river to the lodge at Half Moon Bar. For some reason she trusted me to be the one to take her. We planned to row to Half Moon, in a driftboat I'd borrowed from my boyfriend Harvey—to this day I'm amazed that he lent it to me. He'd just bought it, and it was a beauty: mint green, with black trim and sideboards. It had new ash oars and brass alloy locks. A perfect boat for chasing down salmon and steelhead for sport fishing, Harvey's chosen profession.

Harvey drove us to put-in, and we saw the madrones in the headlights, wet from the weeklong rains.

Michelle asked, "What's that shimmering?"

Harvey leaned close to the windshield and looked up. "You mean the madrone leaves?"

"Yes! That's it—all silver in the lights."

Michelle's eyes glowed like the undersides of the leaves. Her face looked pale and bright, a headlight of its own.

"They're beautiful," she said. "Magical." She turned the bright light of her face toward me. "I see why you spend so much time out here. It's gorgeous!"

When we arrived at the Grave Creek boat ramp, the river looked even higher than the phone report had said. Although the rain had stopped, the feeder creeks and side canyons downstream of the depth gauge in town must still have been pumping in runoff. The three of us stood at river's edge, in our slickers and woolies and milking boots, and watched the muddy water.

"Shit," Harv said. He liked to swear. "It's god damn bigger today than I thought. Are you guys sure about this?"

"No," I replied.

"Why not?" asked Michelle. "Is it too high?"

"I don't know," I said.

Something seemed strange, even besides high water, and Michelle put her finger on it. "There's no one here. Do you think we'll have the river to ourselves?"

"No, we're just early," I said. "Don't you think someone's bound to launch later, Harv?"

"Maybe so."

"Oh, I hope they don't," Michelle said. "I love it here, all wild and empty."

She was so excited about it. And I did expect other boaters to launch that day, clouds or no clouds, because in that country, you never get out if you don't run in bad weather. Besides, we'd promised to meet friends at Half Moon that evening. So I decided to go ahead and run.

We backed the trailer down to the water, though we didn't have to back very far, since the river covered half the ramp. Then we lined the boat into an eddy, loaded a bit of gear for overnighting at the lodge. Finally I kicked off from shore and took the oars, and we whipped into the current like a leaf in the wind. Michelle sat up front, braced on the gunnels with her mittened fingers. We yelled good-bye to Harv, our hands too busy to wave. Our voices no doubt lost anyway.

The river surged down the middle in big sets of rollers. High water covered most of the rocks. From time to time, Michelle looked back at me, smiling and big-eyed. Sometimes surprise side waves hit the boat in the belly or caught the sharp rails at the bottom, a feeling I've never liked much considering how easily those driftboats flip. I struggled like a madwoman to control the boat. But Michelle didn't seem to know it was scary—she just laughed.

"Is it always this exciting?" she asked.

"No, not always."

"But it is today. The river's alive!"

About two miles down from Grave Creek, the rain started again. Soon Michelle noticed all the tributaries pumping in dark water from the sidecanyons. "What's different about the creeks?" she asked. "They're not muddy like the river. They look like tea. Or coffee."

Dark, leaf-stained water poured from gullies and swirled into the foamy water out in the current. With the ground already soaked up the way it was, little canyons normally choked with gooseberry and fir flushed and spilled that darker water. Some of them might not have run that high in years, dark and churning, streaming from places we couldn't see.

The huge falls at Rainie had washed out, just an easy channel of small swirling waves. That was no sweat, although it spooked me to glide over a ledge of rock, normally so tall above the river, buried far below tons of water. At Horseshoe Bend, the water had risen as far up the bank as people say it gets when the gauge reads 20 feet, more than twice as high as had been reported that morning. And in Mule Creek Canyon the water reached clear to the top of the gorge. But I had no way of knowing how high the river was by then—it was higher than I'd ever seen, and higher by far than anyone runs solo. I didn't say anything to Michelle, so of course she got me for being quiet.

"Everything okay, captain?" she asked.

I didn't answer. I was listening. "Shh."

Michelle turned to me, surprised. Then the question left her face, because she heard something, too. Something like a train coming from a long distance, or a thundering jet. Or every last bear out in the woods growling. But it wasn't any of those things. I knew what it was—rapids. Big rapids. Above a certain size they lose their friendlier white noise and start roaring. Maybe they become all water and no air, and the rumble fills the sky and shakes your ears and heart. I knew that the sound meant that downstream, not quite around another bend, was some big whitewater. Blossom Bar in flood.

The startling thing to Michelle was there was nothing to see. Blossom bellowed below us, but just the lip of it showed. High water had filled the rock garden down in the belly of the rapids and covered the marker boulders, so from up top we could only see a curved edge of muddy water rolling ahead. Deceptively smooth, the tongue led into the rapids and then tumbled rocket fast into whitewater.

The only thing to do was pull to shore. I rowed so hard

to get over, I almost overshot a little rock cove I was aiming for in the right bank. We got into the cove, though, and I pulled close to a cliff where we moored the boat. My hands shook as I tied up the bowline, while Michelle waited. Then we kind of crept downstream to scout the rapid.

"Good God," I said when I saw it.

Blossom looked huge and ugly. The usual house-sized boulders were under water, and they made nasty reversals that kicked up froth and spit like I'd never seen before, and haven't seen since, on that river. The standard run, where you swing wide to the left and cut back hard right, was not the usual high-way through. Not nearly so. Rowing the normal route would mean pulling through some ungodly rowdy water and around two deadly sucking holes.

Michelle watched me as I watched the rapids. Finally, she asked, "Why's this so much nastier than everything else?"

"Blossom's got the biggest rocks on the river, for some reason."

She asked the million-dollar question. "Can we do it?"

"Well, I've run stuff this bad before," I said. But I never had in a driftboat, and I never had when there wasn't another boat waiting downstream to bail me out. Which for some reason I didn't tell her.

She kept looking at me, just waiting, I guess.

"We could walk from here," I said. "It's only two miles to the lodge." But we were on the wrong side of the river—we'd have to line the boat back upstream about a half mile before rowing across. The thought of the two of us lining there, with all the fast current and slippery rock, scared me. "Or we could hike out to the road." Ten miles on the river path. We could make it, but well after dark.

"What should we do?" asked Michelle. Her face was bunched up and worried, and I wanted her to stop looking that way. I thought of the driftboat, so pretty, bucking at its mooring upstream. Overnight like that, it might bash up. What would Harv say to that? At least if we made it through the rapids and got to the lodge, there had to be a quieter eddy above Four Mile Canyon.

From the corner of my eye, I could see Michelle's anxious face. "I think I can make it through," I said. "It's just two good pulls."

At that, her face lit up like a full moon after a storm. She tightened the straps on her lifejacket. "Then let's go."

I peeled off my heavy clothes, a habit I'd picked up on trips running big water. But I decided to keep on my rubber boots. Michelle chose not to strip down—she was cold enough already, she said. She just sat on her haunches on shore, one hand holding the bowline, waiting for the signal. I stepped into the driftboat. From my seat at the oars, I listened to Blossom and strained to see my entry. I had to make it across the river, to the tongue of water that led to the safest path through the rapids' spray, shock waves, and foam.

"Okay, Michelle!"

She coiled the rope, and we pushed into the current. I rowed left with everything I had, reaching far ahead of me, finishing each stroke with my legs. Michelle watched downstream, sitting as still as you please. I made the quickest, strongest strokes possible, arching all the way up as I pulled, recovering quickly. Even so we only reached mid-river by the time we'd floated to the top of the rapids. The current was just too fast for me—something I hadn't felt before, not even earlier that day.

As we swept closer to the falls, Michelle got a good look at what lay ahead. She turned and screamed, "Pull, pull!"

The tongue I'd been aiming for was still far to the left. We'd only made it as far as the big pyramid rock, at that water level just a tip of granite in a huge mound of brown water. At the last minute I punted and ran a chute between the pyramid and the gorilla-faced rock. We dropped over a little falls, where the boat made a sickening, scraping sound over something below us. Still, we landed straight. But it was a hard fall that popped the oars from their locks.

"Shit." I slammed the oars back in place.

Michelle still held the gunnels. "We're okay! Pull! Pull!"

I pivoted to the best angle I could and rowed, but the upstream rail of the boat caught in a surge that spun us. We washed out of control toward the Volkswagen rock, just a monster midstream hole at that water level. The driftboat turned nose first into the hole and stopped, but just for an instant before shuddering, snaking up, and twisting over on top of us.

I've always thought that life's like the river, but anybody can see that. There are backwaters and shallows, bridges and dams. There are smooth parts and rough parts. There's the tongue, the smooth "V" pointing to entry for boats, usually followed by rough, bucked-up water in a rapids. The tongue, for all its glassy tranquility, accelerating toward chaos. With all its metaphors of probing, licking, and truth-telling. Once you're on the tongue, you're committed. There are no brakes—you have to ride it out.

In Blossom that day, riding it out meant feeling a wooden boat crash down on my back in the middle of a river in flood.

Being sucked straight down by the water tugging on my boots—staying down so long and tumbling around so much I doubted I'd see daylight again. And swallowing enough water to make myself heave, with no one waiting to pick us up.

Michelle's swim was bad, worse than mine, maybe because she'd kept on her heavy wool clothes. When my boots finally sucked off and I shot to the surface for air, though, she looked fine. Seeing her facing the waves as they came and turning her head to the side to breathe, I felt inspired to do the same. I put aside the panic that had gripped me underwater. But she was a hundred yards downstream, and something told me I had to close the distance and make sure we both got out of the water. By the time I got within twenty yard of her, things had changed. She was stuck in some whirlpools along an eddy fence, wheeling around like a spinner cast out for trout. Floating low, she barely churned the surface with her arms, looking too weak and lost to find shore. Her nose looked red, and her hair hung in her eyes.

I caught the eddy. "Swim, Michelle! This way!"

As I reached for her, she looked around but not far enough to see me. A current shot her back out into the river.

God, I thought, I can't catch her. There's no way I'm strong enough to go back out there.

Then I saw it—one of Harv's oars sliding by, fast and sleek, out in the river. It looked so streamlined going by us that I knew what to do. I took a deep breath and pushed back out into the mainstream. I swam low and hard, telling myself to go like the oar. Far ahead, Michelle's head showed up past dozens of curlers, and I strained and shoved and broke through each wave, never taking the long ride over the tops, but slicing straight through them, like a ten-foot, solid-ash oar.

When I closed in again on Michelle, she was swirling on another eddy fence. This time she floated even lower than before, swallowing water, and got caught in the swift current headed upstream.

"Michelle! No, stop!" I swam for her, afraid she'd wash out the top of the eddy into the river again.

But, no, I caught her. With my right hand, I grabbed for her lifejacket, pulling her back from the whirlpools on the eddy fence. With my left hand, I reached wildly for shore, catching hold of something thin and tough hanging in the water from the muddy bank. It didn't feel like much, but I held to it as if it were a gold-plaited lifeline. As I pulled on it harder, I saw what it was—a tree root, hanging down from the bank and washing along in the water.

I pulled us over and dragged Michelle up the steep bank. We tumbled onto a little patch of slicked-down grass, gasping and coughing. When I looked up, I saw Harv's boat in the water, just across the river. All I could do was stare. The driftboat looked like a big fish come to surface, slowly turning in the current. I had to sit and watch the boat roll out of sight when I wanted so much to run along after it, to do something, anything. But I sat next to Michelle, holding her hand, and didn't move.

"Thanks," she said. "You saved me."

"The hell I did, Michelle. I about killed you."

"No, you saved my life, so let me thank you."

"Okay."

"Thank you."

We stayed the next two nights at Paradise Lodge, just across the river from Half Moon Bar. Willis, Paradise's winter caretaker, had been reading by the fire when I knocked on the door. Still wearing my lifejacket, searching for any shelter, I was amazed to have stumbled onto a neighbor lodge. Michelle followed behind, staggering on the path. Willis's jaw dropped when he answered the door. He said later that with my skin so pale and eyes so frightened, I'd looked like a phantom.

Willis took us in, fed us, gave us dry clothes. For all his hospitality, though, I still felt I'd never be warm again or stop shaking deep inside. We played cards and word games, passing the time as the storm still raged outside, and I felt racked by tremors. I tried to hide them—excused myself to go to the outhouse, hide behind the lodge trying to throw something up, only to return to the fire and tremble some more.

The first night at Paradise, I called Harv on the lodge radiophone.

"We flipped in Blossom," I told him.

"Shit!"

"I lost the boat." Silence. "The last I saw, it was headed for Four Mile Canyon."

"Damn!"

"I'm sorry."

"Hell, don't worry. I mean, shit—"

The radiophone operator interrupted. "Excuse me, sir, this is a citizen radio system. You'll have to watch your language."

"Sorry. Shoot, don't worry about the boat. I'm just glad you guys are all right."

"I guess I shouldn't have run Blossom."

"Maybe not. But hell, you made it out."

Harv never did get mad about the boat, even when we recovered it, bashed and bruised, a week later. Someone had pulled it up on a gravel bar, visible from a road across the river.

On our way home from Half Moon, we spotted the boat sitting out of the water, green and wet in the foggy morning, looking like a ghost come out of the mists. Still, we trusted our eyes, that it was really there. We drove around to get it, loaded it up on the jetboat trailer we were hauling, and returned home both shaken and triumphant.

I never saw Michelle again after that fall. One of the factual things I've written about our trip was that she died not long thereafter. I'm forgetting now just how much time she had, but she was taken by cancer. And although I've told it otherwise, I wasn't there in the end. Whatever choices she made about treatment, how she fought or didn't, what things went through her mind—none of these are subjects I can discuss with authority.

Michelle and I never talked much about the accident, although I wanted to. Those few nights at Paradise Lodge, she'd said that she believed she was going to drown after we flipped. She thought few people would miss her, and she felt okay about giving up—in fact, she admitted that by the time I'd pulled her out of the water, she'd surrendered. Still I was always fishing for something more—maybe validation for running Blossom that day, even though I knew it had been harebrained. After recovering Harv's boat from the gravel bar, I hazarded a question, of Michelle: "Do you regret running it?"

She thought about it. "It might have been the right choice for you," she said. "But from now on, I'm making my own decisions."

Her words have stayed with me. They especially dogged me in the weeks that followed the accident. As I went over my own bad choices again and again, the frightening tape of the river played in my head—the water too high, the speed of the current at Blossom, the helpless sensation of not making the tongue—all of it. The weeks passed, and I chopped wood, taking care not to miss the stump again. Splitting green rounds, dry rounds, junk wood—anything—talking to myself. Going over and over the truth.

Gaining a Loss

Deliquescent, capable of becoming liquid by the absorption of water from the air.

—Dictionary of Geological Terms, Third Edition

Taking the first big step off the beach at Bass Rapids, I hoisted myself onto a ledge of black schist. Down below, someone was singing "Here Comes the Bride." I turned back to look down onto the beach I'd just left. My river colleagues, still life-sized, stood on the sand waving good-bye. I waved back to them, not wanting to leave. Yet wanting to leave. My final trip in ten seasons rafting professionally in Grand Canyon, and I was hiking out of the Inner Gorge halfway through. Scheduled to be married in a week, I had to get up the trail to the South Rim where my fiancé, Matt, waited to drive me to California. We'd be wed in my parents' backyard in the wine country. Then we'd continue to Pennsylvania, where Matt had enrolled in graduate school. All the plans were set.

During my last two seasons of work on the Colorado, I'd been a National Park Service river ranger. A woman in green and gray, running patrols by raft and kayak. In the Canyon, rangers put on the water any time they need to, and naturally the flexible launch dates are sometimes postponed. For some reason I can't recall, my last patrol was being put off later and later. I told my bosses that the delays were fine, but I still had to make August 1st my last day of work to get to the church on time. No problem, my bosses told me, you can hike out at Phantom Ranch. Wait, they said, the water's high. Can you get the boat through the big drops? Stay until below Crystal? We agreed I'd row my boat through the upper Canyon, past Phantom Ranch, and through Crystal Rapids. Then I'd hike out across the river from Bass Camp. A ranger named J.T. had been training at the oars and would row my raft for the rest of the trip.

It had been hard to turn my equipment over to him, to anyone. I'd refrained from reminding him to do the buckles in a certain way on the rocket boxes, to remember to keep the seat pad strapped over the cooler in the hot sun. To ship the oars whenever the boat was at shore so the blades wouldn't chip on rocks in eddies. To remember—never mind. The raft was his. I was leaving the river.

The morning of my hike out, I started at dawn. Everyone

had risen early to see me off. The sun was just hitting the tops of the canyon walls far above us, a strip of orange light on red rock. Embraces and promises all around, and I was off into the dark schist of the Inner Gorge. Picking my way up through the rocks, I found the official Bass Trail, a narrow footpath winding past cairns that marked the route. I'd never walked out the South Bass Trail, but I had a good map. And Matt, who had hiked it before, would be waiting at the top to pick me up.

My companions, now the size of velvet ants, hadn't left camp yet. They were still down below, folding up the tables in camp and loading bags onto the rafts. I finished climbing through the Vishnu Schist, got up into sedimentary rock, and checked again in time to see boats shoving off from shore. They crossed the eddy, then pulled into the main current. Sweeping downstream, their sterns turned away first. My friends were becoming too small to tell apart. Still I knew who was rowing lead, who was bringing up the rear, and that J.T., the novice, was in the middle. In my boat.

My tears didn't start until my friends had rounded the bend at Shinumo and swept out of sight.

In Pennsylvania, I immediately took to the forests. Found solace alone, away from all the strangers that suddenly filled my life. I adored Matt, and I loved being married, but I missed the river as if it had been severed from my body. And as much as I liked the writing job I'd found, I felt the walls too close and my voice too loud for the space. I was constantly bumping up against the limits of my office. Inexplicably feeling like overturning desks, chewing up bookcases, punching down walls to make more room.

In the woods I could breathe. There was something familiar about the sun and shadow on footpaths and the calm feeling among the trees. It wasn't water, but it was good. There were new sounds and sights to learn, like yellowthroats trilling unfamiliar songs and Tennessee warblers hiding in honeysuckle and dogwood. The giant pileated woodpecker pounded its thrilling, primeval beat. I often wandered solo among the oak and maple.

A few months after I'd made the big move, my mother called to say she'd been diagnosed with lung cancer. "I'll be starting treatment this week," she said.

My father spoke from the other line, "Chemotherapy and radiation."

Stunned silence. Then I asked, "What's the prognosis?"

"Well, we don't know yet," she said.

My father added, "The doctor says we'll have to watch the blood markers."

I made a quick decision. "I'm coming home."

My mother: "Oh, no, honey. Stay there with your husband."

My father: "Let's wait and see what happens."

That evening, Matt joined me on my walk in the game-lands behind our home. He held my hand as I stumbled unhappily along one of my normal routes. At a pond where I usually turned home, three pairs of wood ducks lighted on the water. One female called eerily before touching down: is this the right place? Who else is here?

We circled the pond. I tried to follow Matt into a field of shoulder-high corn but caught my T-shirt and jeans on a bramble of wild rose. To hell with all this overgrowth, I thought, fighting my way out. Blood beaded up crimson on a pricked thumb. Why the hell did I ever leave the river? I yelled, "This place sucks!"

Startled, Matt whirled. A pair of deer spooked and bounded past him through the rows of corn. Their white tails flapping in tandem, the deer plunged into the trees beyond the cornfield. The woods closed behind them.

A river basin is a highly organized network of drainages. Small- and intermediate-sized tributaries join to form larger streams, thereby resembling other natural systems with branches or blood vessels, such as trees and animal bodies. Some say the systems evolved as they did to best transport fluids, disperse nutrients, and absorb energy to their cores. The headwaters of tributaries pull runoff from the hills, feed it to a main stream. Tiny capillaries of the human circulatory system transport blood outward to exchange oxygen at the skin surface. The limbs of trees reach to find light for life-giving photosynthesis. In each system, the extension of the individual arms brings energy and power.

In trees, particularly my mother's favorite live oaks, boughs and branches stray in their own directions. Botanists have a term for the baroque divergence of tree limbs, as I learned from reading one of my mother's natural-history books. In it she'd underlined *deliquescent*, a hiss of a word meaning dividing repeatedly. Branches separate and separate again until they're wisps of wood barely related to the oak's central trunk. "You wouldn't think it," said my mother, "but the tree really needs all that branching." The divisions look primitive, but they're fundamental to the tree's growing and maturing. More complex endeavors than these are hard to find.

In Pennsylvania, I had a karate weapon—a bong, made from an oak tree. Matt and I harvested the young oak from the Pennsylvania woods. Before we cut it, the tree stood among other saplings of oak and birch, crowded together so closely that they guided each other's growth straight up, the only direction they could go. We prayed for the spirit of the tree, then we felled it and dragged it home. Night gathered. In our porchlight, we counted eleven rings in the cambium. Eleven years old—a good age. Too young to have developed a crooked spine, too early to deliquesce.

Over the course of several months, I skinned and cured the oak, turning it regularly to prevent warping. The bong grew lighter upon drying. Being a former tree, it was not symmetrical end to end, so I whittled the wide parts and smoothed off knots with a drawknife. When the tree had cured thoroughly, I sanded the rough spots and rubbed the entire length with linseed oil to keep the wood from cracking. Finally, the bong seemed about as finished as it would ever be.

Too aware of its imperfection, I felt embarrassed to take my weapon to karate class. It was not nearly so straight or balanced as the bongs of my fellow students, who'd ordered theirs from a karate-supply house in Philadelphia. Mine measured more than the standard inch in diameter, had a crook where a branch had grown, and hadn't been machine sanded. But Matt, who had also trained in karate, argued in favor of the bong. "I like that it's not perfect," he said. "At least show it to your teacher."

When I took the bong to the dojang for practice, my karate instructor called me out onto the mat. He circled me, scrutinizing, and asked to heft my bong. I handed it over. He studied it—the bumps I was unable to smooth, the dark traces of cambium still on it. Then he spun it. He jumped and struck an imaginary enemy. Standing in place, he traced figure eights in the air and ran the bong through two quick forms. Turning to me with a smile, he handed back the bong. "It may be different," he said. "But that's what makes it yours." He looked meaningfully at the other students.

I visited my mother in California, where she was spending a lot of time dozing in the backyard at home.

"Mom!" I said. "You look great."

"People keep saying that—as if I shouldn't."

"What've you got there?" I pointed to the I.V. bag.

"A prep for chemotherapy."

She took the treatment, her books and letters beside her. I read pages of my journal, jogging my memory about the river so I could write about it one day. I tried to remember the Canyon. The adrenalin at Lava Falls, Crystal, House Rock. The green-blue waters of Havasu Creek and little travertine-rimmed pools to dive in. Matkatamiba, Deer Creek, Thunder River, Elves

Chasm, Tapeats Creek. Even the names too beautiful to believe. Mudbaths in the Little Colorado. Flash floods pouring red water from Soap Creek, Ryder Canyon, Nankoweap. Trout longer than your arm in the eddy at Saddle Canyon. The balmy air at night when the sun-baked wind had finally cooled enough to allow sleep.

A breeze brought me back to my parents' backyard. The boughs of their pear tree swayed out on the lawn. My mother stirred.

"I've been asleep," she said. "It makes me wonder." She closed her eyes again.

"Wonder?"

She blinked awake. "Oh, yes. I wonder if I'll sleep away the last little time I have left."

"No! You're just tired from the treatment."

"What good is it? I can't eat, can't walk much. This isn't living."

"You'll get back to doing the things you love."

"You think so," she said. She lay back to sleep.

Returning to Pennsylvania, I took my sorrow to the dojang. There, among the karate weapons, I found that I was not only sad but angry. The big punching bag became the enemy—God and nature. I railed against a world that would allow my mother to get so sick just as she was starting to reap the benefits of a life of hard work. I saw her death as a neon sign advertising the unfairness of life. Life is unfair, I ranted. It's unfair, it's unfair, it's unfair. I attacked the bag with punches, flying side-kicks, spinning back kicks. As I lashed out in a blind fury, nothing felt so good as spinning my oak bong and striking targets with it until my arms ached.

My karate instructor advised me to find equanimity. "Channel your rage," he said. "Learn to manage it, or it'll consume you."

I tried. Still my anger not only ate at me but also spilled into other parts of my life. Startled, my friends argued that I'd hurt myself somehow. My husband began to draw away as my rage took on a life of its own, out of control. I felt more of the discomfort I'd discovered when I left the river—how much I wanted to bust out every day, overturn things if I had to, live large.

I had to keep moving. With the money I'd saved from my final river paycheck, I bought a new mountain bike to ride through the gamelands. Feeling no qualms about going alone, I pedaled furiously on the gravel and dirt roads looping generously through woods and fields. I made a game of finding how fast I could go without falling.

First I'd climb the rutted utility road near the corn, standing on the bike pedals to make headway. At the top of a hill, I'd heft my bike over the barbed wire near the university dairy buildings, coast the grade to Sweet Creek, pass the weather station on top of Ag Hill, and dip into the deeper woods where sprinklers sprayed treated sewage into the undergrowth for a research project. The wet trees smelled ripe, like compost.

Retracing my route full speed, I started to relax and look around. Branches of white oak swayed above me. I wasn't even going full speed when gravel crunched and gave way beneath my tires. Braking hard, I knew immediately that it was a mistake. My handlebars cocked toward the road, and I tumbled head first over them, my eyes on the dirt fast approaching. Slamming shoulder first, I slid, and the bike tangled with my feet as it careened off the road. I landed face up to the sky, my eyes on tree limbs high overhead. Again I wondered, what the hell am I doing here? If this bike were a damn boat, I wouldn't be having this trouble.

I didn't move until I heard the chug-chug of a vehicle climbing the road from Sweet Creek. Remembering that men traveled the roads in pickups to unclog choked pipes and sprinkler heads, I pushed myself to the side of the road. Found myself right beside a pink balloon, largely deflated and come to rest on the dirt shoulder. *If you find this,* read an attached note, *please write and tell me where you work and what you do. Sincerely, Charlotte W., Boalsburg Elementary School, PA.* I stuffed the note in my pocket.

A burly, gray-haired maintenance man in coveralls stopped to help me up. He lifted my bike into the back of his pick-up and gave me a ride home. After I'd cleaned the gravel from my wounds, I hobbled to my desk. Suddenly energized, I wrote, *Dear Charlotte, I found your balloon and message out in the gamelands. More than two miles from Boalsburg. I'm a writer and hope someday to be a mother, too.*

Another trip to California. My father picked me up at the small municipal airport near home. "You won't recognize her," he warned. He didn't say much more than that. Mom had spent a week at a local hospital and had begged to leave. If she was going to sleep away the rest of her life, she at least wanted to be home. My father had made arrangements for a home care nurse.

I found Mom lying in her bedroom, near a picture window onto the backyard. Hooked by drip system to morphine and compazine, she was resting as the light changed outside. Shadows of fruit trees moved and swayed; tendrils of wisteria blew in the afternoon breeze.

There was a keen look on my mother's face when she awoke. The color of her eyes had changed since my last visit, from brown to cateye green. "Mom? Everything okay?" I leaned to kiss her, found her scent strong and acrid.

"Oh, sweetie, I knew you'd get here. . .." She closed her eyes. Her legs barely mounded a light blanket. Dark blue veins showed in her hands; needle marks colored her skin. On her nightstand stood a vase of golden rattlesnake grass and a small plastic basin.

My father headed back to the airport for a short business trip, the only one he'd dared go on in many months. He left me with contact numbers. "Call me immediately if anything changes," he said and slipped out while my mother lay sleeping. When she seemed to be out for the night, I collapsed onto an unmade bed in the guest room, pulled up a comforter, and instantly dozed off.

Near dawn, my mother's vomiting started. I ran in to hold her basin. "It's okay, Mom." I wiped her face with a damp cloth. "Where's the nurse's chart for your pills?"

We had a number of prescription drugs, some that would help with the nausea, but I couldn't find the administration chart on my mother's nightstand. "Do you know where it is?" She shook her head and moaned. A deep, frightening sound rose from her throat. I held the plastic pan as she hunched over again, this time racked by dry heaves, tears building. She moaned again, the sound built to a scream, and she closed her eyes and bared her teeth. I remembered a trapped-rabbit scream I'd heard once, out in Utah. It had sounded neither animal nor human, as my mother sounded now. Her blind pain was followed by sheer fatigue. She collapsed back on her pillow.

Over the next week, with the nurse's help, we tamed the nausea, but at the price of even more of Mom's waking hours. As she lay sleeping each day, I sat at her bedside, going through pages of the river journals that I was sure would be essential to some future pieces of writing. "Worked ten years in the Canyon. 1976 to 1984 as a guide for Arizona Raft Adventures. Then two years as a ranger. 1985 and 1986." I hardly recalled—it seemed so far from the bedroom where my mother lay dying.

Whenever the nurse came, I helped her turn my mother, change the sheets as they were stripped, and launder everything. I brought ice chips and held them to Mom's lips. In the evenings, I sat beside the bed as she waved her arms in her sleep. When she awoke and saw the puzzled look on my face, she said she'd been painting away the pain in her back. Looking embarrassed, she fell back to sleep.

The night before my planned flight back to Pennsylvania, Mom looked bright eyed and at ease. She asked to be propped up. My father had returned, and we sat listening to the radio at my mother's bedside.

"Is there anything you want?" she asked me. Her face glowed like hot coals.

"Want?"

"Yes. Have you gone through my jewelry box? Or talked to your father about furniture?"

My father gazed forlornly at his plate.

"No!" I said. "Mom—I'm thinking of staying for another week."

"No, honey. You're needed in Pennsylvania. You can come back later. . ."

"Later?"

"Your life is back there." She closed her eyes. "Why am I so tired?"

I told her I'd be back soon, but she didn't answer. Her face appeared serene and quiet.

Another use of *deliquescent* is not directly botanical—it means dissolving or melting away by absorbing moisture from the air. Fungi deliquesce in moist woods, as do the petals on flowers. In cultures where deceased loved ones are dispatched in boats on rivers, the skin, hair, and vital organs of the corpse deliquesce. Become liquid through contact with the air.

To depart a loved one, whatever the cause, is also a form of deliquescence. Divorced spouses or deceased parents deliquesce, as certainly as ice melts in water, as they simultaneously never leave us.

A few days after my return to Pennsylvania, my father called. "It's over," he said. "She's past the pain." I flew to California one more time. Friends drove in from long distances, some of whom I hadn't seen in decades. There were childhood friends from the old neighborhood, my best friend's parents, my mother's hiking group and their spouses. We gathered in the funeral home on the main street of the small town my parents had recently chosen for their retirement. As our family entered and took our places in the first rows of the funeral home, I glimpsed the sorrowful faces lined up farther back.

One of my mother's closest friends delivered a tribute, in which she told of some hours spent with my mother during her last few weeks on earth. This friend had walked my frail mother out of her bed at home, settled her under a blanket in the back seat of a van, and driven to some of their favorite places. On their drive they'd spoken of many things, and at one point they'd passed under an arbor of oaks that arched the road.

My mother had remarked, "The trees look beautiful today." She'd seemed happy as they'd driven under the oaks that leaned to touch each other across the road.

"I'll never again pass under trees arching overhead without thinking of that drive," said my mother's friend. "And I'll always remember Mary."

Back at my father's house, our closest friends gathered to continue the services. We listened to my mother's favorite music, as we stood around unable to eat all the wonderful food brought in by neighbors. Gifts and cards poured in, one a note delivered with a bunch of white, helium-filled balloons. The note asked us to release them in memory of my mother.

My father carried the balloons outside and let them go. They disappeared into the blue October sky. First the balloons looked life size, then the size of ants. Soon they were too small to tell apart. How far would these ones travel? One mile, two, three—never mind. It only mattered that we'd let them go.

Back in Pennsylvania, I returned to a house of sympathy. My karate class had sent roses, my coworkers had hand-delivered a bowl of fruit. An avalanche of cards covered the kitchen table, but I'd read them later. My first thought was to get out to the gamelands. Matt joined me, holding my hand as we strolled among the glorious reds and golds of the forest. Beneath oak limbs with crimson leaves tearing like wind-blown flame, I stopped to breathe great lungs full of air. I figured, here is the reason I left the river. It's home, and there's room to breathe. Wouldn't my mother love the exquisite clarity of such a day?

Reading Water

Even where the channel is straight it is usual for the thalweg, or line of maximum depth, to wander back and forth from near one bank to the other.

—Fluvial Processes in Geomorphology

A hawk soars into view. It blows in on a breeze from over the ridge and above sunlit treetops high overhead. Dark face, brown wingtips and trailing edges, red chest and shoulders, fanned and sharply banded tail—it's a broad-winged hawk. I sit up to watch as it drifts out of sight, quiet as a cloud. Then I lie back in the long grass, savoring the early morning cool settled on Guthrie Creek. It's April at Galbraith Gap in central Pennsylvania. By afternoon each day this week, temperatures have risen in a pressing humidity. But here by the water, the day seems perfect. Overhead, birch and ash trees weave and sway, a dappled arbor.

About twenty feet away, my friend Drew stands at the side of the country road that follows the creek, shaking his head over his mountain bike. He has a problem with his rear hub. Sorts through his roll of wrenches and curses. The tools he needs have gone in my car over Galbraith Mountain with my husband, Matt, and our friend Paul, who are shuttling vehicles for our bike ride. Drew can't believe his bad luck. If he wants to make the ride, he'll have to drive the approximately 40-mile round trip to retrieve his tools when the guys return from the shuttle.

As Drew ponders his predicament, we hear the purr of an engine far up the mountain. This isn't a road that gets much traffic—it must be Paul and Matt. But could they be back so soon? The vehicle descends the mountain's sharp switchbacks, muffled by the full growth of forest, hardwoods tiered with fir beyond the creek. Sitting up, I get a glimpse of a station wagon rounding the road's last bend. The car, a red Volvo, cruises into view, then pulls up alongside us as if our spot by the creek were a scheduled stop.

The Volvo is at least thirty years old, with glittering clean, original paint. Out steps a lithe man with a full head of gray hair and open, friendly face. "Name's Larry," he says. "Need a hand?" Drew raises an eyebrow and shakes his head hopelessly. He thanks Larry anyway but says he needs special mountain bike wrenches. Larry bends near the bike and peers at the problem. Straightens up and returns to his car.

Instead of waving good-bye, as we expect, Larry opens the rear of his vehicle to reveal a hefty toolbox. He promptly pulls out a perfect set of bicycle spanners and goes to work on Drew's problem hub. Whistling to himself all the while, he mentions that he owns the bike store in town and lives upstairs, above it. He twirls the spanners, makes adjustments, and in three minutes, Drew's bike is repaired and ready to ride.

As Larry packs up his tools, we offer to pay him, but he flatly refuses. He closes up the rear of his Volvo, waves cheerfully, and drives away.

Drew asks, "Did that really happen?"

"I think so. I saw it, too."

Larry has disappeared down the road around a gentle bend of creek. The forest closes behind the hum of his car.

"Maybe he was a seraphim," I say, "one of those temporary angels sent to earth."

"Oh, like, 'Larry the bike store owner? He's been dead for years!'" Drew gapes down the road, silent now but for the murmur of water. Maples hung with spring leaves and winged seeds arch the road. Drew shrugs and spins his bike's rear wheel.

When Paul and Matt return, we ask if they passed anyone on their way over the mountain. No, they say, the roads were empty the entire way.

When I first river-guided professionally, I ran one- and two-day trips back-to-back nearly every day during the summer. The Stanislaus, my home river, had a small hydroelectric dam upstream of our run. The water in the river ran high when demand for power peaked in the Central Valley—say during working hours, when offices needed air conditioning. Other times the river ran so low it seemed as if someone had shut off a faucet upstream.

The water-level fluctuations, both daily and seasonal, gave us regular lessons in how the river varied depending on flow. The *thalweg*, or deepest or best navigable channel, didn't always follow a direct path. On one key day early in my training, I followed a boatman friend named John through the long, straight, placid reach of the Stanislaus below Razorback Rapids. As I rowed down the middle of the river, choosing the course where the main flow had been weeks before, I noticed John's boat meandering from one side of the river to the other. He kept his hands on the oars but barely exerted himself, simply using the oars to adjust his boat's position on the water surface. He moved briskly downstream through the calms with little effort. Even as I rowed steadily to keep up, he beat me by finding the strongest flow and doing the bare minimum to stay on it.

"It's true," John told me later. "You've just got to use the current. It'll carry you if you don't fight it."

Our mountain bike ride takes us to the fire tower on Galbraith Mountain. Pedaling ahead, Matt, Drew, and Paul arrive at the top before me. By the time I ride up, still in a full sweat, Paul and Drew have climbed the steep lookout stairs to the tower deck. Matt has waited for me at the base, and when I arrive, he takes my hand so we can climb the steps together. Halfway up, we pass Paul and Drew on their way back to the bottom.

"There are blueberries down there!" says Paul, as he and Drew fly down the steps.

Matt and I continue upward, to the deck and locked-up tower cabin. In moments we are gazing out over the valley to the next long, wooded ridge thirty miles distant. The stultifying heat, which would have been upon us any other day this week, has not come today. In fact, a spring breeze chills the lookout, and Matt shivers. He's had time to cool off since the ride up.

"I don't have a jacket," he says. "I'm heading down."

"Wait. I may have something for you." I open my knapsack and pull out a large sweater. It's an old cast-off of Matt's, patched at the elbows. A sweater I love to wear. "Hmm—I don't remember packing this, but here."

Matt slips on the sweater and stays a while. Below us, silos and century-old barns dot fields of corn and alfalfa. Creek courses unspool like green thread over fields and between farms. The moment is perfect, as is my place in it. A rare feeling for me, who only months ago left my long-time guiding job at the Grand Canyon. I walked out of a river trip at the halfway point, to get to my wedding on time, and moved across the country to a strange green landscape and community where I knew no one.

Drew and Paul are down below, stuffing their mouths with blueberries from a scraggly mountaintop patch. Soon the guys are gagging, making faces. "They looked ripe!" They reel away from the blueberry plants and spit out the acid taste.

Matt chuckles and thanks me for bringing the sweater. "You saved me. I'd be down there choking, too, instead of enjoying the view with you."

"That was lucky," I reply. "But don't thank me—thank the spirit who packed your sweater."

One summer, I introduced the concept of following the thalweg to a boating friend who'd struggled with rafting for years. She'd spent most of her river time wearing herself out in the flat stretches. Her raft had regularly landed on sleepers, rocks lurking just under the surface in riffles, because she couldn't read the rumpled water above them. Invariably she'd tremble above rapids, knowing they'd kick her ass every time.

I sat behind her, coaching as she rowed through some of the smaller rapids in the Grand Canyon. Pointing out that she was working too hard, I urged her to look carefully, find the main current, let it carry her downstream. I told her about reading water. "Let the river pull you to the best channel. Look hard, you'll see it. Sooner or later you'll just feel it." In flatwater the thalweg will carry a boat downstream faster than rowing. In rapids, it will find a way between boulders or through the deepest, safest channel nine times out of ten.

She looked near tears. "I've tried to learn from so many boatmen, even a few women. *Nobody* ever told me I was working too hard."

I agreed. "I know. If you're in the wrong place, you can bust your butt and tear up a lot of boats for nothing."

In biking down Galbraith Mountain, we discover a quick descent—a steep, single-track route over a scrub-covered slope on the mountain's drier side. It's definitely a black diamond run, to borrow a term from resort skiers: deeply rutted, and covered with sharp boulders and deadly patches of gravel. Drew volunteers to make a test run. He lowers his seat for stability and takes off.

The rest of us watch, holding our collective breath, as Drew makes his way down, braking cautiously. In places he skids dangerously or has to get off to walk his bike. A few minutes later, he's worked his way to the bottom. He stands far off the trail to get out of the way of the next to follow.

Paul decides to go next. He's either forgotten his helmet or doesn't own one, which bothers us far more than it bothers him. Looking over the trail briefly, he shoves off over the edge. At first he's in control, feet on the pedals, hands working the brakes, looking fine as we cheer him on. Then he reaches the midway point and things go wrong. He fends off the trail's banked sides, his feet sticking like outriggers off the pedals, as he careens first one way and then the other. He twists in a rut, caroms off a boulder, shoots off toward the brush. I should be screaming with horror at his wild ride, but Paul looks so comical, I laugh so hard I can't breathe.

Miraculously he returns to the trail. One last bump bounces him clear off his seat, but he grips the handlebars anyway, stays with his bike, and lands again perfectly seated. He reaches the bottom and, carried by momentum, keeps on going. Watching from below, Drew suddenly realizes that Paul's not about to stop, lifts his own bike with one hand, and scrambles wide-eyed for the brush. Paul sails full speed past Drew, finally skidding and wobbling to a stop hundreds of feet farther on.

Amazed at Paul's good luck, Matt and I pick our way

to the bottom. Down at the run-out, Drew sits on the ground, dressing a puncture wound in Paul's calf. Paul thinks a small branch must have poked his leg. His ride also ripped a sole from one tennis shoe, which flops open at the toe.

"That was easy," says Paul, hoisting a small roll of duct tape from Drew's repair kit. "Not bad at all."

Adopting a strategy of following the river thalweg worked well for me during many years of guiding. Figuring that I probably had more brains than brawn anyway, I decided I'd better use them both in their true proportion. Let the river do the physical work. Whenever new initiates asked to pilot my boat, they jumped to the oars enthusiastically, figuring they'd get on the rowing seat and engage in a fluid dance similar to the one they'd just witnessed. As smooth as a heron in flight. Instead they discovered how heavy the boat was and how unwieldy the oars. They cried, "But you made it look so easy!"

I did because I'd learned from the river. Somehow, somewhere, I'd come to understand the language of water. Sometimes that just meant staying on the thalweg: at low water it meant meandering through a riverbed and at high water surging straight downstream. My boat's momentum almost always came from river power.

Life off the river seemed much harder. After retiring from professional guiding life and driving to Pennsylvania, I found myself in uncharted waters: marriage; indoor jobs, ice storms in the winter. I needed a new map, not just to negotiate roads and trails where there'd once been stream channel, but to help me navigate a landscape where "go with the flow" seemed to no longer apply. To get down a river, we had to find the deepest channel, with some exceptions. To start a new life, what were the rules?

Farther down Galbraith Mountain, we bike down an old logging right-of-way along a falling ridgetop. Sweet, young grass covers the road. The perfect, easy grade leads through patches of shadow and light, past the blurred woods. I pedal with caution, mindful of surprises in the grass, especially hidden rocks that Matt calls "cowards." As I ride prudently, Matt and Drew bike easily just ahead, shoulder to shoulder, as Paul cruises far ahead and out of view. The distance grows between me and the others, and the farther ahead they get, the greater my fear of cowards. I grip the brakes. Soon I'm burning with frustration at the foolish fear that slows me down.

I round a bend near a creek, sweep into the forest and the riparian cool, and nearly collide with Matt and Drew. They've stopped on the road in the shade of a spruce. Matt grins at me. "We were missing you," he whispers.

"Yeah." Drew picks leaves and twigs from my rear brake, his voice also a whisper.

"Why are we whispering?" I whisper.

Matt points toward the creek. A slicked-down beaver busily strips leaves from a willow sapling. He's holding the willow in his front feet, chewing with a soft chuck-chuck, the sapling quivering as he munches. His absorption is so deep, he could hardly be less aware of us if we were behind glass.

Then it hits me—I've been here before. The three companions, the bend in the road sweeping into the forest, the beaver working the creek—I've seen them all somewhere. Although this is the first time I've bicycled with Paul and Drew, lived in Pennsylvania, or descended Galbraith Mountain, finding the beaver and standing at the creek both seem so familiar they feel scripted. It's like watching the beaver in Jones Hole Creek, Utah, or seeing otter deep in play on the Salmon River, Idaho.

As *déjà vu* swirls around me, Paul's voice comes from somewhere down the grade. "Hey! Where are you guys?" The beaver slaps the water and dives. Cursing the unseen Paul and his noisy ways, Drew and Matt cruise after him. I'm left standing by the creek, wondering about the puzzle of *déjà vu*, the phenomenon that some say means being in the right place at the right time.

Watching the surface of the water, I try to discern movement that will show me where the beaver's gone. I pick out a little current in the creek. Gentle but direct, it follows the deepest part of the channel—the thalweg—and makes a beeline for the languid, glassy pool where the beaver disappeared. As I stare, getting fixed on it, the beaver pops up in the current, catches it, floats back to his place of work. He doesn't seem worried about the temporary upheaval in his world. Shrugging off my daze, I pedal to catch the others. The breeze that chilled us up top catches our backs, a tiny breath of summer. Before our rolling tires, the new grass rolls out like a carpet, and the cowards seem less scary. I lay off the brakes, so we're coasting, carefree. I marvel at my good fortune to have found such a place and time. The going is easy down Galbraith Mountain.

Traveling Bodies

A kinematic wave is a concentration of units, particles, or individuals, but through which these units can move. A concentration of cars on a highway is an example.

—River Channel Bars and Dunes

Behind me on Highway 50 in the Sierra Nevada, cars string like silver beads on a shining necklace. I'm pushing the speed limit on this winding stretch of road, but traffic bunches up behind me as if I were parked. I pull off to let the cars pass. Traffic or no, I'm not going to rush on this beautiful Saturday morning in October. To my right, the South Fork of the American River plummets downslope in giddy counterpoint to the traffic's uphill flow. The forest above is filled with bronze and gold, and the year's first snowfall drapes the slopes in a fluffy cloak. Last night's storm has moved on, leaving crisp, unseasonably cold air that rushes through my open window. It would be criminal to hurry past such a day.

As rivers often do, the American calls me to look closer. Exiting to the right, I coast down the hill and into a streamside campground. Parking my car, glad to be rid of it, I walk to the water's edge. There I enter the river's world, a galaxy away from the highway cosmos I've just left. The whispers of riffles and cries of rapids cover the growl of traffic filtering down from the road. A water ouzel peeps as it wings upstream, careening two inches above the water. Upriver, a lone camper kneels on a sandbar to dip fresh water into a cooking pot. The peaceful scene calms me, and I forget the press of cars that has pushed me down here.

This stretch of river changed greatly last year. Summer fires charred the slopes and banks bordering the American, and winter floods not only washed away the fire debris but also scoured clean some parts of the riverbed. In other years in October, branches would be overhanging the river, clutching faded leaves, but this year no trees crowd the banks. A few alders lie stripped and flattened, mere skeletons. Most vegetation has been uprooted and carried off, leaving the river channel a bare bed of rock. A geological, not biological, showcase.

There's plenty to observe here. I move close to a cluster of boulders that breaks up the river's smooth flow. The boulders are chiefly granodiorite, Sierran bedrock eroded from the mountain and carried downstream by the river. I step out and onto the largest one, a gleaming hunk of white rock dotted with black and pink minerals. In the boulder's wake lies a collection of other sedimentary particles of various types, materials, and sizes: at least seven smaller boulders queue up in two lines behind the big rock, like children at a lunch counter. An infilling of cobbles and sand grains sparkles in tiny eddies between the boulders. A fistful of pine needles crowns the entire assemblage.

Taken together, the particles form a congregation, which has alliances I seek to untangle. Which boulder arrived here first, dropping to the riverbed to become pioneer of the current? Who dropped in behind, once the first had fallen and altered the current? Not one of the rocks speaks, at least not to me. They simply crowd together, mute and secretive, as gossips huddle until it's safe to talk.

In the Grand Canyon, one of my fellow boatmen was Phil, a natural leader, an ex-Green Beret, and a terrific salesman. I'd met Phil on my first-ever river trip, when I was seventeen years old, and he seemed too handsome to be real. He was tall, tan, and beautifully lean, with a white, even smile. As lead boatman on the trip, he'd oozed self-confidence in an easy way, and the people had become admirers: women with sunscreen ("Phil, do my back"), men flattering his youth and glamour ("Phil, you're terrific"), boatmen deferring to him ("Ask Phil. He'll know").

Still his military training and experience in Vietnam were right at the surface. He ran a tight ship and brooked no slackers. Smoking cigarettes during his rare free moments, he mingled with the people in a perfunctory way—professional and thorough, but detached.

Years later, after I'd begun working trips in the Canyon, I saw Phil again at the boathouse in Flagstaff. He'd relaxed a great deal, and his face wore a perpetual grin.

"Phil," I said, "you've changed. You look so happy."

"I've taken up meditating," he said. "You should try it."

"But I'm already happy."

"Yeah, but you'd be *really* happy."

Phil was sold on meditation. He'd sit cross-legged mornings and evenings on river trips, finding privacy on his raft, his sleeping tarp, at the water's edge. Off the river, we'd see him sitting in the driver's seat of his Volkswagen van, as if asleep but upright, ignoring our amused dance outside to get his attention. He stayed with his practice wherever he went, religiously pursuing enlightenment. Soon he became trained to teach the techniques of Transcendental Meditation, and he offered classes everywhere. He would lecture to little knots of passengers at lovely stopovers in the Canyon, like Redwall Cavern, Matkatamiba Canyon, and Deer Creek. On beaches, at lunch spots, the people sat transfixed before him, watching his beauty and listening to his words.

Phil would teach entire crews of boatmen to meditate during our fourteen-day river trips, earning himself the nickname Guru. I wasn't one of the boatmen who learned TM on the river, because in those days I felt too green to take on another distraction. Still, I wanted to learn to meditate, because I wanted to be *really* happy. I asked Guru to teach me TM at the end of the season. He did, leading a small group of guides in a September training at a friend's home. In the gathering fall days, among the yellowing aspen near the San Francisco Peaks, Guru taught us not only the philosophy of meditating but also its nuts-and-bolts practice. We listened to him lecture, learned the meditator's vocabulary, and studied charts on improving our nervous systems through practice. Near the end of the training, in private, quiet rituals, Guru gave us our mantras. We latched onto them gratefully, as a craftsman seizes a missing tool, and took every opportunity to sit together in practice. For years, we had worked, eaten, and slept together, followed each other through the big drops, and fished each other out of the river when we flipped. After studying with Phil, we could also deepen our connections by joining together in the vast, uncharted realm of meditative consciousness.

Standing in the American River, I study the aggregate of sedimentary particles at my feet. Big keystone boulder, smaller boulders lined up behind, cobbles and sand in the interstices, sprinkling of pine needles. Squinting my eyes until the individual particles blur together, I think of river processes. I remember studies of river movement conducted from the perspective of *kinematics*, named for the Greek *kinēma* (motion), the same root that gives us *cinema* (moving pictures). Kinematics analyzes motion from perspectives other than those of mass and force. The interaction of individual moving particles becomes important, say, or the effect of environment on the traveling bodies.

Taking a kinematics approach, researchers identified a class of waves that are physically distinct from the classical waves in other systems. Most waves travel from Point A to B: ocean waves arising offshore and running onto the beach; "The Wave" traveling through a crowd of sports fans; a quick pulse of tossed wash water surging toward a drain. In contrast, kinematic waves stand in place. By physical law, individual particles traveling in the same linear direction bunch up, the slower ones impeding the progress of the faster ones like poorly shot pool balls. Made of massed-together water molecules, kinematic waves take shape where the river channel constricts, or the water strikes a variation in the stream bottom, or a tributary flash floods into the main channel. The crowding of particles manifests as general rises in river level or as waves maintaining a static position in the channel. When the kinematic waves assume a classic symmetry, they're known as standing waves, glass-faced, often perfect dunes of water.

Researchers extrapolate on the kinematic wave theory by applying it to other groupings of like particles. Gatherings of drinkers in pubs, bunched-up automobiles on Highway 50—both are examples of kinematic waves. To qualify, the groupings only need be started by a single, exceptional individual. And though they may look static, particles come and go. Water molecules, cobbles in a cluster, grains of sand shifting on the riverbed—all move ahead even as the form of the wave remains.

Meditation is often regarded as a solitary practice. After all, the Buddha reached enlightenment by sitting alone for forty-nine days under a fig tree in Bodh-Gaya. Don't the silence and stillness of meditation add up to hermitage? In truth, practicing with others deepens the experience, as I discovered while meditating at Phil's TM training. Sitting in a room, our breath measured and quiet, we shared a sense of deep rest. In years to come, when we'd learned to integrate our spiritual pursuits with our guiding jobs, we'd gather when we could in little pods on river beaches. There we'd check out among the sounds and smells of the Canyon.

These days I meet with a mindfulness meditation group in my community. We begin by sitting with eyes closed. Later we embark on a period of walking meditation. I imagine that as we walk we look like zombies, moving languidly in a garden with tangled tomato vines and pepper plants, overgrowth of lavender, and sage abuzz with insects. We are not so far off from my Grand Canyon *sangha* of years ago, and, in moments of meditative transcendence, I find myself back there. Then, Phil was our keystone, our exceptional individual. Today it is someone new. Tomorrow it may be me.

On the American, I decide that the largest boulder at my feet was the first to come to rest on the stream bottom. The others no doubt subsequently fell into place behind it. Among the boulders and cobbles is wedged a chunk of driftwood, which, when wiggled, easily dislodges and moves downstream.

As the driftwood shifts, the assemblage of particles changes—a sand-laden current jets through the wave, breaking up and carrying off the collection of pine needles.

My attention travels downcurrent with the pine needles, gradually widening in scope to take in the whole stream—a mountain river of stunning beauty, studded with sunlit boulders, reflecting clouds scudding in a remarkably clear sky. Higher water will move the bigger pieces, and subsequent waning flow will carry in replacements. On this particular bend in the river, the kinematic wave will gather again. The streambed is configured perfectly for it. If I return in a week, I'll see that the group has commuted. It will look different, but new particles will fill in for those that moved on. Although the wave will have changed, it will remain the same.

Confluence

Tributary, any stream that contributes water to another stream.

—Dictionary of Geological Terms, Third Edition

My good friend Louise writes that she won't be able to attend The Race this year. Instead she'll be touring in Europe. "But *you* must go," she says. "I'm not sure it can be done without us." And I'm not sure I want to go without her, even though the annual Tsunami Ranger and Wild Sea Gypsy Race near Half Moon Bay is an event not to be missed. I've known Louise for about three decades, and she remains unique—she's slender and striking, with long, flowing black hair and flashing dark eyes. Her skin stays tan year-round from the sun and her body strong from her job as a Grand Canyon river guide. Her good looks are outstripped only by an entertaining wit and native intelligence that spark fun, fast-paced conversation. At Half Moon Bay, her presence would just enhance the event.

Louise and I guided together in the Canyon and have shared other experiences on the water—time spent kayaking near her long-time home in southern Colorado, hikes on Pacific beaches, a paddling trip down Baja California's Sea of Cortez. In Baja, we met Michael and Dennis, a couple of highly skilled ocean kayakers who live on and paddle the northern California coast. Michael's home on a wave-whipped sea cliff is headquarters to the Tsunami Rangers, a band of extreme-condition boaters with a chain of command modeled loosely on that of the U.S. Navy. They kayak plenty of hairball water, all while maintaining their clearcut rank and file. Maintaining order not only keeps them afloat even in the most severe conditions but also supports their claim that they have absolutely no death wish.

At the San Francisco airport, I identified Michael and Dennis right away as possible kindred spirits. Michael led them both along, confidently lugging his outdoor gear and the pack of cameras he uses for his job as an adventure photojournalist. Dennis followed behind with a sort of bemused expression. Although we didn't speak, we traveled parallel schedules—they changed planes when I did in Los Angeles, stood behind me when I joined Louise in line for the flight to Mexico, and passed through customs just ahead of us in Loreto. It was becoming obvious that our paths were meant to cross. So when Louise and I met up with our sea kayaking group in the tiny Loreto airport, it came as no surprise that Michael and Dennis stood among them. We four faced each other with a spark of something that could only be called recognition.

An obvious fact about the joining of two rivers is that it changes them both. A muddy river clouds a clear-flowing stream; a larger stream subsumes a smaller one. A tributary can dilute the receiving stream, raise or lower pH, elevate levels of dissolved oxygen, and add suspended sediment. No river that joins another remains unaltered.

Not even the river's name stays the same below a stream junction. The larger or longer upstream arm gives its title to the combined river downstream, and the lesser upstream arm assumes the status of tributary. Sometimes it's tough to discern which of the two upstream rivers should be relegated to tributary—they may appear to be fundamentally the same in terms of length, watershed area, channel size, and amount of discharge. In that case, the downstream river is dubbed for the upstream arm most closely aligned with the downstream system. So the dominant river below the junction continues in generally the same direction it's been flowing.

The name for stream unions has been fixed forever: *confluence*, which comes to us intact from sixteenth-century

Latin: *confluentia*, or "the act of flowing together." A fittingly fluid label for the joining of waters. I've always figured that if river runners met annually, much as the historic mountain men and fur traders convened for rendezvous, we'd call our gathering Confluence. We'd come together from all over the world to trade confidences and share advice about rapids and river levels. There would be gear swapping and tale telling late into the night. At Confluence, although you might arrive as an individual, single and alone, you'd soon mix it up a bit. You'd become integrated into something bigger and fuller, larger than the sum of the parts.

May 2nd breaks cloudy and windy. Louise and I have agreed that at least one of us must get to The Race, so I drive to Montara Beach, north of Half Moon Bay. The sky at the coast is bright but sunless because of a cover of ocean fog, and the gray surf stacks up on shore. When I arrive at Montara, I see two dozen sea kayaks already lined up on the beach. The kayakers have suited up in full-body wetsuits, booties, gloves, helmets, and lifejackets, and they're either making last-minute adjustments to their boats or standing at the water's edge sizing up the waves. Entrants need not necessarily be official Tsunami Rangers, but they must be expert kayakers.

When I find Michael and Dennis at the lineup, we greet

each other as if we parted only last week, although we haven't seen each other or talked in over a year. I notice that they've added a distinguishing feature to their usual paddling gear—a set of viking horns attached to each of their helmets. Most of us know *Vikings* as a group of pirate Norsemen who plundered European coasts in the eighth to tenth centuries, but *viking* (with a small "v") is also Scandinavian for *sea rover*. The horns serve as a reminder of their dedication to exploring things *del mar*.

"Will you be trying hard to win?" I ask.

They laugh. "We don't need to be first," replies Dennis.

"No," Michael adds, "just viking proud."

At the base of the sea bluffs, racer orientation begins. Most of the initial speech is a witty but informed enumeration of Dos and Don'ts: *Do* stop to rescue another racer in need or you'll be disqualified ("We have rescue boats," says Soares, an officer in the Tsunami Rangers. "But we don't know where they are"). *Do* circle Flat Rock en route along the approximately five-mile course to the finish. ("You can't miss Flat Rock. It's a rock. And it's flat.") *Do* pass by Phallic Rock ("You'll recognize the shape") and Mavericks Beach ("If the waves are up at Mavericks, it'll be hairy"). *Don't* stop until you've pulled your boat all the way over the finish line at Miramar. Many races have been won in the final dozen yards by the contender who's quickest to run up the beach.

Then Michael steps up in his modern-day viking garb

to read an inaugural poem. It's his own composition, epic and powerful. His words bless the paddlers and wish for the wind and waves to remain at their backs. He invokes Neptune, Roman god of the sea. He wishes each boater safe passage to the finish at Tsunami headquarters. A moment of respectful silence is followed by an announcement of a five-minute countdown to the starting signal. All we can do now is wait.

Maybe everyone has her favorite stream confluences. I know I have mine. I love the junction of the Middle Fork and Main Salmon Rivers, not only because they expand upon meeting and slip into a greater valley, but because people congregate there. I met a long-time love at the confluence when he drove shuttle on my first-ever Middle Fork trip. Over the next few years, we met there again and again, whenever either of us ended a trip. The one who'd been landbound drove to meet the other at take-out with a full heart and great anticipation. The confluence is a place of many greetings.

Another splendid confluence is that of the Green and Colorado, where the olive-green water of one river mixes with the brick-red of the other along an endless eddy fence. The combined rivers then wheel around a tamarisk-fringed point bar before descending into the wild water of Cataract Canyon. The confluence sandbar has been the scene of many late-night parties and of encounters at high noon when everything that breathes slips into scarce shade. In high-water years, the sandbar has also been the meeting place of hushed boatmen full of anticipation of the mammoth rapids waiting downstream. The confluence is a place of much expectation.

My favorite confluence of all, though, is that of the Yampa and Green in Echo Park. A sand island has built up several feet tall where the rivers meet. In high water, the Yampa shortcuts to the Green at the island's north end; in low water, the Yampa takes its time and swings all the way downstream of the island. At any level, the rivers mix and flow together around Steamboat Rock, a massive peninsula of sandstone with walls and ledges favored by nesting peregrine falcons and swallows. The wall rises cool and imposing above passing rafts, a cruise ship dwarfing its lifeboats. Pockets in the rock return echoes to even private conversations.

Across the river from Steamboat Rock lies Echo Park, with its grove of cottonwood trees marking the National Park Service campground. A friend of mine rangered there for many years, his backcountry station a little cabin a hundred yards from the river. He had the job of maintaining order in the campground, as well as riding herd on boaters as they floated past Echo. Legends of his work are legion, not the least among them being tales of his fondness for nude sunbathing by the river. On one occasion when he was enjoying his riverside basking, a group of boaters floated out of the serenity and privacy of Lodore Canyon and to the base of Steamboat Rock. No doubt their party had fallen under the river's spell, come to believe they were alone in a wilderness, and sighed at relief at having reached the confluence safely. They stripped off their lifejackets and lay back on their rafts.

My ranger friend looked up from his sandbar, noticed the boaters floating unlawfully without personal floatation devices, and jumped to his feet. He demanded that they put on their lifejackets.

They gaped at him in all his splendor and asked, "Who are you?"

He slapped on his green National Park Service ballcap and replied, "I'm the ranger! Put on your lifejackets!" Apparently they immediately complied.

The confluence is a place of countless surprises.

Each morning on our kayaking trip in Baja, Dennis was among the first to rise. He's a strong paddler and extremely fit but with an admitted addiction to cigarettes. Before sunrise, he prowled the beach, enjoying an early smoke, always in a watch cap and camouflage suit that looked government issue. My first impression was that he was familiar with guns—perhaps a hunter, a survivalist, or a lifelong Marine. But as I got to know him and found none of my speculations true, I had to wonder how camo fit his lifestyle.

After drinking enough *cerveza* one night, Louise and I dared to ask him about his outfit. He confessed that he'd been a nurse in the military. Said the suit is lightweight, of high quality, and invisible to infrared sensing. "And I had to pay for the camo with my own money, so I'm going to wear it, damn it." He claimed that Michael had invited him along so there'd be at least one *pasahero* on the trip amenable to being photographed. And Dennis earned his keep—he posed for picture after picture, paddle in hand, on land or in the water.

Another question had been troubling me. "How on earth do you rangers patrol for *tsunamis*?" The name for great sea waves produced by submarine earthquakes or volcanic eruptions, *tsunami* is Japanese—*tsu* for harbor, *nami* for wave—meaning that the waves can overwhelm even your most sheltered areas. Tsunamis move ferociously fast (up to about six hundred miles per hour). With low observable wave heights on the open sea, they're hard to see as they approach the continent.

They pile up to one hundred feet or higher only when they enter shallow water, and then they can sweep clean a coastline. Curious villagers have been killed in great numbers when they've rushed to shore to watch the strange sight that often precedes tsunamis—the sea sucking away from shore before it regroups and charges inland.

"Do you figure you'll see the tsunamis coming?" I wondered. "Do you stand all-night watch from a protected overlook in Half Moon Bay?"

"No, no," Dennis assured me. "It's nothing like that. *Tsunami Rangers* is just a name."

Just a name—and a good one—implying that if the Big Wave comes, Dennis and Michael and all their buddies will be ready for it in whatever craft is at hand.

At the announcement of a second five-minute countdown, the boaters in The Race laugh nervously. They try out their cockpits, adjust spray skirts, stand in the surf, and size up the breakers they'll be entering. One racer who's arrived late is dashing around in a frenzy looking for the sign-in sheet. He seems to find it, because in a moment he takes his place among the racers. Minutes later, there is a third five-minute countdown.

We chuckle but figure this is really it. The crowd backs off from the racers to allow them room to move.

Then it's the final backward count from ten to one and they're off. The boaters charge down the sand toward the water, kayaks and paddles in hand. They hit the water at a run, push their boats seaward, and take their seats. A few skilled and lucky boaters skim directly through a window in the four-foot wave sets, Michael and Dennis among them. No doubt they make it because of their own consummate sense of timing—they reach the surf, pause long enough to let a breaking wave pass, shove their boat down the wave's seaward side, jump on, and paddle madly toward the next incoming wave. They crest it before it breaks on them, hitting the only hole in the surf that's to come for many minutes. Within thirty seconds, they've moved out through the window and into position behind the handful of boats ahead of them.

The shore behind Michael and Dennis is now littered with boats. Those kayakers who didn't slip through the sets are strewn like driftwood along the beach. The boater who couldn't sign in is among them. So is Soares, the Tsunami Ranger officer, and he isn't about to give up.

One older fiberglass boat has cracked in attempting the breakers, and its owner, followed by his dog, is the first to shrug and drag his boat back up the beach. Other racers keep trying—they charge the surf, gain a little headway, begin to build momentum, and then are stopped short by the breaking waves that surf them back to shore. After dozens of tries, many drop out. They're simply too weary to paddle the rest of the course.

Michael and Dennis have told me that in past years, they've been among those stuck on shore, and indeed I've seen them in similar circumstances. In Baja one morning, *El Norte* whipped the Sea of Cortez into a choppy surface too roiled by whitecaps to paddle. I'd heard of the dangerous winds, which can drive the sea to a frenzy and swallow entire boats without a trace. *El Norte* can blow up in an instant, the result of distant, changing pressure systems. If it catches you onshore, you can be pinned to a beach for days.

Michael and Dennis volunteered to test the *El Norte* waters by taking a double kayak out past the surf zone. They launched in a relatively quiet spot and headed for the break a few dozen feet offshore. Building up momentum, they topped the breakers, one after another, two, three, four, until they were almost home free. Then one of the last big ones dumped them. Their double immediately swung parallel to the surf. After a few attempts at rolling, they were out of the boat, each ranger holding to one end of the kayak. A breaker grabbed the double and surfed it and the rangers back toward shore at about twenty-five miles an hour. Michael and Dennis staggered in and admitted, "It's big out there."

Hours later, when *El Norte* had tapered off, we all had a chance to try the wild surf. We took our turns launching those who felt ready, watching them paddle out briskly, like wind-up toys with arms circling. Soon it was my turn to shove out into the wild surf in my single, and I was off, paddling for the outside some two hundred yards out, giving it everything, reaching for all the water in the ocean to pull it behind me. I heard the cheers from shore as I made progress, then Dennis's voice shouting, "Rudder!" Thanking him silently, I released the rudder so I could steer out through the breakers.

Making for an opening in the surf, I faced the waves head on, still paddling, never stopping, never letting down my guard. Then it was fierce paddling, not forgetting what had happened to the rangers. I caught some air on the backsides of breakers, slowed by some caps but not stopped. Each wave came in like a blue hill to be climbed, up one slope to its apex, then down into the long limb on the backside to the next dark trough. Each wave seemed to obscure not just the sky but the universe. Within minutes the sea had smoothed out into huge swells, and I realized I'd made it outside the surf zone. I sat like a duck in some storm-riled pond with the others, hushed and waiting for the last kayaks to come from shore.

"We did it," said Louise. "We don't know squat about paddling in the ocean, but we made it."

I nodded. "All that time spent on rivers counts for something."

"It's kind of the same. The people you meet, the water. Kind of different and kind of the same."

Later, when we'd reached our evening's destination, the rangers congratulated those of us who don't sea kayak regularly.

"That was tough paddling," said Michael.

Dennis nodded. "We're impressed."

"Good," I replied. "We've been wanting to impress you guys all trip, and we've finally done it."

For all their heart-stopping excitement, the waves we encountered in Baja couldn't have competed with the surf today at Montara Beach. Some of the kayakers who've been pinned to shore make it outside the surf—Soares gets into the running after eight tries. As for the rest of the boaters, they're pushing out, tipping over, rolling back up, washing ashore, floundering exhausted in the foam. I'm reminded of the many painful hours I spent over several years shivering in a river kayak, trying to learn to roll, getting dumped in even minor rapids, gasping for breath. Hard to reconcile such miserable times but for the many hours of pleasurable experiences kayaking has brought me as well.

Team by team, the paddlers power past Phallic Rock. They circle Flat Rock, skirt outside the famous surfing waves offshore of Mavericks Beach, and paddle the long haul past the Half Moon Bay marina. Team by team, they arrive at the final beach, surf the small breakers in to shore, and pull themselves out of their cockpits as fast as their numbed hands will allow. Their final effort is to run up the beach past photographers and the handful of staff who are timing The Race, pass the colorful flags marking the finish line, and collapse over a small sandy slope on the far side.

The crowd cheers the racers through this last leg of their journey. We urge them to keep going, you're almost there, don't give up, run, run. One racer who's sauntering up the beach shoves his boat aside and tells us all to relax. He never does cross the finish line.

And he's not alone. Only sixteen of the twenty-three entrants actually complete The Race. Most of the non-finishers are those who never made it off Montara Beach. The first-place winners paddled the course in one hour and eight minutes. Our friends the horned vikings appear to have come in fifth or sixth.

Soares finished somewhere near the back of the pack, but he seems unworried. I encounter him later dressed for the after-race party in full gypsy attire, with black Zorro-like hat and flowing cape. "You look fantastic," I say. He grins at these words, bestowed on him by this utter stranger who forgets to take his photograph to show Louise later.

Driving away after saying my good-byes, I'm filled with a happy glow. Today the world seems to have room for kayakers in general and iconoclasts such as the Tsunami Rangers in particular. As the distance grows between me and Half Moon Bay, I'm reminded of a question Louise asked when we first left our friends at the Sea of Cortez. "I feel so vulnerable out in the big world!" she'd said. "How do we go on without the Tsunami Rangers?"

Hopefully we won't have to. We're boaters, joined in something bigger and fuller, more than the sum of our parts.

Life on the Floodplain

Water in excess of channel capacity must flow overbank, inundating parts of all of the floodplain.

—Fluvial Processes in Geomorphology

Valentine's Day, nine months into my marital separation. I pass out flowers at work: camellias plucked fresh from my garden. They're still wet from weeks of heavy rain. My co-workers return from the bathrooms or coffee machine and squeal in delight at the floral surprises on their desks. The data processor in the next cubicle murmurs, "How wonderful." It's the day's sole gift, so I keep one of the camellias for myself. There will be no phone calls or flowers from my husband, who loves another.

The camellia stands in a company mug near a framed photograph of my daughter, Rose. Rose, true blossom, gift of my life.

Outside the plate-glass office windows, rain has flooded the marsh—brimming with ducks and herons, fluttering with gulls, risen to a lake. This week Pacific storms have swept relentlessly onto these coastal hills. Radio news has warned of potential flooding on the Russian River to the north.

For the moment the downpours have stopped, and steam wisps from clusters of oak and bay on the hills surrounding the wetlands. After work, I walk outside, in the washed-clean air. A muddy road leads into pockets of forest behind the marsh, and I follow it. If I walk fast, get my heart pumping, will it feel better? Will it finally stop hurting after all these months? The scent of bay envelops me. Live oaks drip rainwater from curved, glossy leaves; grandfather's beards cloak the bare black oaks. I can barely face them, the arboreal reminders that I will never again take my husband's hand and explore the woods. I feel that I could dissipate into the wet and silence.

On the drive home, the six o'clock radio news reports that the Russian River is close to going overbank. Residents in Guerneville and Cloverdale, river towns, have evacuated their homes in case of flooding. They've had to move temporarily to higher ground.

Seems they do that every year. Annoyed, I switch off the radio.

My husband moved from our home while I was on a river trip in the Grand Canyon. Before I left for Arizona, he informed me that we'd be separating. By then it was too late for me to change my travel plans and stay home to patch things up. I asked him to wait, at least until I returned. He said okay, but the look on his face told me otherwise.

The southbound train arrived, and I stepped up to it. My husband followed, holding Rose in his arms, keeping her between us. She reached for me.

"I can't go," I said.

"Go," he said. "You need a break."

Everything I cared about in the world stood on that platform, and yet when the train pulled up, I got on it. Down the length of California, across the growing miles of desert, I prayed that he wouldn't leave. That he wouldn't take Rose. Scenes of great beauty and wild desolation passed unnoticed. Who cared that forests of Joshua trees stood in ghostly beauty in full moonlight? Who cared about the first glimpse of yellow pine outside Williams? Usually I did, but not now. My only thought was to reach a telephone in Flagstaff to call home.

Disembarking the train in the rush of icy wind funnelling down Mount Humphreys, I shivered and dialed the telephone outside the Flagstaff station. My husband answered. "I've decided we should separate permanently," he said.

"You're talking the 'D' word?"

"Yes. You're too angry. We fight too much."

"What about Rose?"

"I'm doing the best thing for Rose." Silence. "Here. I'll put her on."

Too young at the time to talk much, Rose breathed into the receiver. I chatted to her, trying to sound cheery. She listened a while, then surrendered the phone to her father, who said good-bye.

I hung up and stumbled along the street to the Weatherford Cafe to meet my friends for the river trip, reminding myself to breathe. It was morning at 7,000 feet, and the mountain air felt sharp and cold.

On the river, I ached for Rose and the family I'd worked so hard to put together. The Canyon had always felt like home, but now my heart was back with my daughter and husband. I felt trapped between rock walls, the gorges too narrow, the Colorado just something to get over with. The days seemed interminable, the nights never ending. My river friends, appalled at my depression, gave me all kinds of advice. One said, "You've done all you can. Give it up to the universe."

Another said, "If he doesn't love you, you don't want him, do you?"

And finally: "It's not your fault. Half of all marriages end in divorce."

To that, I replied, "I know. But mine wasn't supposed to."

On returning home from Arizona, I found our house had been emptied with my husband's scientific precision. No sign of him remained save empty spaces. He'd sorted through drawers of silver, boxes of books, the shelves of our garage. Had made his own decisions about what to take—his tools, books, camping gear. His French stoneware plates. Worse, he'd driven away with Rose, although he left a note saying he'd bring her back as soon as I returned.

Immediately I dialed the number on the note. My husband responded by stopping by to drop off Rose. He barely looked at me. When I tried to follow him, to ask him to reconsider, he told me to stay home and take care of our daughter. Then he drove off in someone else's car. *Her* car?

With Rose in my arms, pacing among the rooms of the stripped-down home, I thought, this is too fast. I can face anything but this.

I strapped Rose to my back and fled to the forests. We descended into gullies of dogwood, willow, big-leaf maple, alder. Rose picked leaves, favoring the tough, curled coast live oak leaf. We wandered the vineyards together, she poking the plump clusters of grapes, laughing as they swung to her touch. She tore a handful of bright green grape leaves from a vine and clung to them. Darkness gathered; she flung down her leaves and clung to me instead. Finally she asked, "Home?"

I had to face it and stood again at the threshold of my house, key in hand. Once inside, I played the radio and television simultaneously to fill the void with sound. The country radio ran its usual program of songs about lost love.

> It's gonna take a lot of river
> To keep this broken heart afloat.
> It's gonna take a lot of river,
> Running all the live-long day.

I turned down the mournful lyrics and paid attention instead to the television news. The Russian River had gone overbank. Lives had been lost to landslides, homes buried in mud. In some neighborhoods in Rio Vista, water had ponded ten feet deep. People had fired up their motorboats to try to salvage property at their homes. Flotillas of men wearing black slickers had turned out to rescue animals stranded on rooftops.

Why the hell are those people living on the floodplain? I wondered. I turned off the television, fastened Rose in her high chair. Fed her dinner, put her to bed, snuggled with her when she couldn't get to sleep. Later I rose again to put away food, clean dishes, wipe counters. Finally I lay out clothes for Rose and me for the next day—me to go to work, Rose to go to day-care. Then I fell into bed exhausted, with just enough energy to think, This is how it will be, me and Rose. Good thing the river could never go overbank here . . .

Nine months later, on Valentine's Day. I'm halfway through my after-work commute when the rain starts again. With the radio off, water pelts the roof of my car with fresh ferocity. I'm in no rush to get home. Rose has been off with her father for the week, and my house will undoubtedly be cold on this gray evening.

By the time I reach my neighborhood, the way in is blocked. Water streams from the hills, feeding the floodwaters that have risen in my town. Runoff has filled ditches and surged into intersections, jamming up cars whose drivers line up to ford low spots. With heavy flashlights and weary gestures, patrolmen in raincoats and plastic-covered hats direct traffic. I wait my turn, but it's forever in coming.

When I finally arrive at my own street, I find it's flooded. The storm drains have overflowed, gutters spilling into yards, up front paths. Current a half-foot deep sheets west-east over asphalt, seeking some outlet that isn't already full. As I pull into my drive, its through water up to my wheel wells. Runoff streams into my backyard, over my garden, against my front porch, pushing finally against the sandbags my neighbors have packed at my garage door.

The next morning, I wade out of my street and hike to the hills. I want to see the creek where it plunges over my favorite, secret waterfall. Following a deer trail, slicked down and wet, I bushwhack to the banks of the swollen creek. Upstream the falls crash, in a hidden glade I learned about from my mother. She showed it to me in a dry year, when the falls only trickled over shale. Long before I was married. Now the waterfall roars, plummeting a hundred yards. The creek below is muddy and rushing, smelling of earth, running willy-nilly toward town.

I thought I'd never see my own creek in such chaos, but here it is. The whole mess has gone overbank, and the rescue work still lies ahead. Maybe we all end up in the path of the floodwaters sooner or later, even those of us who think we live on higher ground. Given the right circumstances, even a minor creek pressed between banks of blackberry and buckeye can become a lot of river. I have no doubt now that it could rise even more, remind us who is boss.

In the morning, I will travel south, to the home of my former love. It will be time to retrieve Rose, who has spent a week with him. To get her, I must cross plenty of bridges—over my creek, over the river to the west, over the bay to the south. The floods may still be up, the drive arduous, the destination reminiscent of incidents I'd rather forget. But I'll go all the same, the journey for my daughter the thing keeping me afloat these days.

Of Cobbles, Zen, and River Gods

Imbricate structure, characterized by
thin flat pebbles all tilted in the same direction,
their flat sides dipping upstream.

—Dictionary of Geological Terms, Third Edition

I wade up Lagunitas Creek, naming names. Gray and brown pebbles: quartzite. Russet boulders with conchoidal fracture, threaded with white veins: jasper laced with quartz. Vesicular cobbles, midnight black: scoriaceous basalt. The jasper and basalt I know, but the quartzite stumps me at first, although its tiny, mostly rounded grains say it might be a fine-grained sandstone. Its even texture and inconspicuous structure suggest trachyte, a volcanic rock, but its lack of porosity says massive quartzite. Intrigued, I trek back downstream, looking for cobbles with more obvious clues. I find none and settle on a name for the rock: quartzite.

Again heading upstream, I see dark green ledges of

smooth, runnelled rock: greenstone. Narrow ellipses of mid-stream gravel, colonized by willow: islands in the braided creek. Lenses of silt in the creek bank: migrating stream deposits. Occasional pockets of stones overlapping as neat and regular as aligned and fallen dominoes: imbricated cobbles.

Most of the names I learned while studying earth sciences and examining fluvial environments in the field, but the imbricated cobbles I discovered while running rivers. Imbricated, from the Latin word *imbrex* (a curved roof tile), and meaning covered with or composed of scales or scale-like parts overlapping like shingles. All akin to the Greek *ombros*, rain, a root word suitably water related to describe river-deposited stones. The plumes of a peacock's tail in full show are imbricated. So are the squama of fish, the petals of carnations, the scales on the cone of a Douglas fir.

Rows of mountains in wilderness have been called imbricated. In *Cold Mountain*, Charles Frazier writes that the native people of Tennessee and North Carolina had a word for it: "Cataloochee, the Cherokee word was, meaning waves of mountains fading to the horizon." Portrayed by fine artists as increasingly gauzy with distance, the imbricated landscape looks heaped and telescoped. Layers appear to be stacked one against the other, with no gaps lying between.

My notes from fluvial studies conducted over twenty years ago brim with the names of geologic structures caused by moving water. Once scientists recognize them in modern streambeds, the structures can be distinguished in the fossil records of sediments deposited by ancient rivers. I recorded page upon page of information about them, thinking the notes would help me when I went out looking for signs of water in old rocks. In my notes I constructed handwritten charts with column headings—NAME, DESCRIPTION, OCCURRENCE, SIGNIFICANCE. Rows are filled with lists of erosional phenomena such as scour holes, fluted steps, grooves, rills, dendritic marks, and armored mud-balls. There are names of depositional structures, too, those caused when sediments settle out of streams with waning flow. Their names include sinuous ripple marks, complex interference patterns, streaming lineation, and imbrication.

The names make my head spin, but one thing stands out—the chart row for imbrication includes a tiny sketch of shingled cobbles next to arrows indicating the inferred direction of streamflow. The cobbles angle upstream, the imagined current pressing against the last cobble's smooth face, forming an inviting resting place for the next water-transported cobble to wiggle into position. Thus the stones rest in the lowest-energy arrangement, least prone to upset by a cobble flipping out of sequence.

SIGNIFICANCE: indicates channel conditions. That is, the cobbles settle in imbricated configurations in the active part of the stream. They won't be carried off to the side, say, with silt washing laterally overbank during flooding.

A margin note among the pages reads, "observed in strong eddies in the Salmon River." While guiding on the Salmon in Idaho, I sometimes wore a pair of binoculars hanging around my neck and resting over my lifejacket. When I saw rockbars with coarse gravel fabric in the river's huge eddies, I'd ship my oars and lift the binoculars to glass the cobbles. In the eddies, the backcurrent traveled upstream as fast as we moved down, and the huge rockbars contained almost nothing but cobble-sized stones shingled against the reverse current. Many times no binoculars were required—we skimmed past the rockbars with a foot or so to spare. The imbrication looked stunning—ovoid cobbles as large as Frisbees aligned not only in rows but columns, two or three layers deep and who knows how many more below the water's surface.

We could also gaze down through the clear water at flawless beds of imbrication on the streambed passing beneath our rafts. We were riding above at about five miles an hour, feeling as though we moved twice as fast.

Over the years, my river passengers often commented that the landscapes we traveled made them feel insignificant. Some said it as if they'd never felt that way before. Others remarked that they felt a strong presence of God in river canyons. No sight or vista inspired these theological observations more often than imbricated cobbles, which appeared to have been placed by a deity's hand. God playing dominoes, rummy tiles, mah-jongg. The order of stones followed a pattern determined by divine intelligence, a plan, according to which everything happens as it should. Perhaps, even, God had gone before us to level the rough places and smooth the way for our journey. According to the Psalms, *Great peace have those who love your law; nothing can make them stumble*. In this spiritual scheme, the fabric of imbrication is a piece of God's plan, geology according to divine right order.

Zen Buddhists have another take on life. They speak of *nirvana*, defined as the end of suffering, the abandonment of selfhood. The Buddha taught that nirvana can be achieved by those following the Noble Eightfold Path, which includes the practice of right mindfulness. Mindfulness is the aware, balanced acceptance of the present. Some interpreters of the Buddha's teachings consider right mindfulness to be an essential ingredient of happiness. Clinging to the past brings suffering; anticipating the future brings striving.

Right mindfulness is good practice in the study of rivers. If boatmen are mindful, they can see the aspect and alignment of cobbles in the current just as they occur, whether they dip upstream or down, whether their long axes lie parallel or transverse to the current. Show a Grand Canyon boatman a picture of imbricated cobbles and he will know just what you're talking about. He may not know the name for them, but he can tell you where to find them, for instance down around River Mile 208. The rocks will be there in familiar configuration, if they haven't moved since the last trip through, and everything is just as it is, in the moment. There is no motive to cobble placement, no interpretation, no deification. There are no stories to make up.

During the 1983 flooding in Grand Canyon, I stood with the reverent boatman and renegade Wesley Smith above Hance Rapids, overlooking the huge hydraulics we were faced with navigating. The water was higher than we'd seen in many seasons on the river. Hance looked big and ugly, with no clear path through. Upstream, the reservoir called Lake Powell had more than filled, and Glen Canyon Dam seemed shaky to those of us living our lives downstream. Chunks of the dam's own concrete had washed through the spillways. As boatmen, we felt both thrilled and terrified, not knowing whether the dam would burst and we'd be riding the Big One to the Gulf of California. Probably we'd just continue to experience the newest form of water torture, in which the river rose in increments every day but no one could predict when it would stop.

Observing the rapids at Hance, Wesley said, "We'll have to offer extra prayers to the river gods to let us through."

Although frightened of the big water, I disagreed. "No," I said. "You just have to put your boat in the right place." Wesley turned to me with a huge grin, delighted that I presumed our fates were in our own hands.

We proceeded to run the rapids. It did no good to imagine the familiar river before the high water. We had to plunge into it as it was in the present, mindful of the markers onshore and on the surface of the water, foreign as they looked at the new water level. At any one moment, there was only a wall of water rising before us, or the sudden rush of sunlight and air between waves—where we caught our breaths—or the river pouring in on us, threatening to sink us. We must have put our boats in the right place, because we survived. Whether due to right mindfulness, divine right order, or the benevolence of the river gods, we never knew.

I advance farther up Lagunitas Creek, plowing the water's surface with my thighs, mindful as a monk. The surrounding woods are lush with redwood, sorrel, periwinkle, and azalea. The shallows reflect light shows onto overhanging branches. In the dappled sunlight, water striders touch off concentric ripples. Beneath my feet the creekbed feels particularly lovely—fine-grained sediments have been winnowed by the stream, leaving an aggregate pebble surface.

I hear the rattle of a kingfisher behind me, and I follow its call back downstream, retracing my path. In a moment I see the flash of wings, but not before I pass a small community of imbricated cobbles in the bank. They are stones of quartzite, jasper, and basalt, the rock types I've seen earlier. They lie cheek to jowl, long axes horizontal to some ancient current, planar faces dipping upflow to the water in the long-gone ancestral stream. Now they dwell in the moment with me, cobbles without a story, stones without a motive, pantiles set without benefit of roofer.

Faith in the Dry Season

Ephemeral stream, whose channel is

at all times above the water table.

—Dictionary of Geological Terms, Third Edition

Late summer going into fall is when I set out to the creeks looking for clues. By autumn in this valley, the streams have dried up like wrung sponges, their bare beds testimony to something lost. I must believe I can find traces in disappeared pools and past currents, because I get up and walk the bare creek bottoms. They draw me from home like a migrating bird that rises and moves with assuredness to the foothills north of town. There, stream canyons funnel like unsigned roadways into the flanks of old, worn-down mountains.

Sometimes I go at dawn, when parched leaves falling from the neighbor's birch remind me it's been months since the rains. Then as I walk, I see stars fade in a strip of light building over the hills to the east. Other times I go in the evening, after a breeze has curled through my kitchen screen with a cool edge that says summer's heat is breaking. By the time I return home, the stars are visible again, as if they'd never disappeared, and didn't I just know they'd be back?

This morning I walked to the hills before dawn, and now I'm standing in a tributary to my hometown creek. This nameless drainage winds out of the mountains, dodges under bridges, and twists past vineyards and houses, joining the mainstem creek on its north side, at the edge of some country properties. About four feet deep and three wide, this tributary channel today lies thick with everything but water—an orange carpet of fallen buckeye leaves, handfuls of wild plums. A flicker's barred feathers crown a bank of cobbles topped with deflated algal mats.

Spring floods are just a memory. In March this creek danced with standing waves, smelled of earth, and rose to within inches of its bridges. Heady days, when weather reports promised more, though the channel couldn't hold it. Glory days, but they've passed. Now here it is, the end of summer going into fall, and this dry bed won't see water again until the winter rains return. It runs only during big storms. By definition it's ephemeral—from the Greek *ephemeros*, meaning, "lasting one day only." With no way to rewet its surface year round, the ephemeral stream is dry more days of the year than not.

Today, evidence of wet storm pulses shows up only in clues that say something's passed. Tangled tree roots adorn undercut banks, the soil that once held them gone with the high water. Isolated pockets of dead pollywogs lie with their black bodies drawn into question marks. Crawdads scutter over the streambed, digging for moisture at the fringes of sandbars. In the drying mud, cracks join end to end to form polygons whose edges curl in the sun.

There is water farther down and deep here. I know it's there because of the skin of dry mud that arches up from the underlying wet layers, as if pulling away from an unloved twin. If not for the telltale curling of the mud, would we know about the moisture below? Would we recall high water and have faith in the dry season? If not, the aridity of the place might overwhelm us, days without rivers stretching endlessly ahead.

Late summer going into fall speaks to others as well, especially my friends in the river-guiding community. These are boatmen who ran or still run rivers commercially throughout the West. Year after year, through their twenties and thirties, and sometimes into their forties or fifties, they have matched their lives to the change of seasons. Although I left the business years ago, I still feel the shift in my cells: in spring, blood runs higher than water. The river calls us to get back to it, to jump in and hold on. Summers bring kinship, good food, and rapids to jar the senses with adrenaline. The weather is so easy we live on sandbars under balmy skies and stars so numerous we have no need of nightlights.

Then the summer ends in fall, a restless time. By autumn in desert river country such as the Grand Canyon, a chill lingers in the air long after dawn. Trips taper off, and boatmen find themselves more and more in town. These former bronzed gods shuttle from car to laundromat carrying mountains of clothes that must be washed several times to come clean. Those who have no winter jobs or schools to attend hang out in coffee shops and bars, no longer surrounded by the passengers who praised their skills and worshipped their heroism. Sometimes the guides hustle to get scheduled on off-season trips, if not in the Canyon, then in Costa Rica, Chile, or New Zealand. Most often, though, the boatmen simply pass through an amorphous period of transition, during which they run idle errands while making decisions about the coming winter.

An indifferent, isolated time, fall is also when we've lost many of the boatmen who have taken their own lives. Our community mourns them still, although in some cases decades have passed. A surprising number of skillful, daring men have met their end that way—enough so you could call it occupational hazard. The death toll for guides by suicide may outstrip even the numbers dead by drowning. Sometimes they've used the bottle, sometimes more immediate, violent means, often carried out in the collapsing days and fading light of autumn.

Whale was a mountain of a man, soft spoken, utterly competent. I never knew him by any other name, never considered calling him anything else. He worked the Grand Canyon every season since 1970, as a commercial guide on motor-powered pontoon rigs. After nearly ten years of seeing him cruise past our oar-powered trips, I finally met him one lunchtime in 1983. He walked into our river camp from his rig parked far down the beach. Swampers, green boatmen twenty years his junior, flanked him on either side. With their sunburnt hair and mirror shades, they might have made a formidable entourage had Whale not pulled off his sunglasses and grinned.

I pumped his arm and introduced myself. "I'm the one you pulled out of Crystal last trip."

Whale smiled and nodded.

"Thanks for saving my life," I said.

He shrugged, and our conversation went no further. Wondering why he didn't speak, I peered at him, with his boyish, tumbled-blond boatman's hair and freckled, pink suntan. He just squinted and made soft little assenting sounds, as if "Ah, shucks" were lodged somewhere deep inside him. We stood staring at each other a few more moments, and I became aware of his complicated eyes—they had distance in them that seemed to reflect much more than the 224-mile stretch of Colorado River between Lee's Ferry and Diamond Creek. Then he and his swampers, their curiosity about me apparently satisfied, retreated back to his rig.

To say that he'd saved my life may have been no overstatement. He'd fished me out of the river a few weeks before, after I'd flipped an eighteen-foot-raft in high-water Crystal Rapids. Crystal is an unforgiving place—waves the size of ocean swells, house-sized reversals that eat boats, and a raft-ripping rock island lurking only dozens of yards downstream. In ten years of working full time as a commercial guide in the Canyon, Crystal was the only place I'd even come close to witnessing a drowning. There, I saw more than a few nasty flips and more than one boater who couldn't get out of the water. A handful had drowned in the whitewater-choked gorge downstream. Crystal was one of a few rapids I never wanted to run, much less swim, no matter how many Canyon trips I had under my belt.

After my own flip in Crystal's main hole, I found myself in the same speeding current that had claimed others' lives. My boat foundered somewhere upstream of me, too far away to be of help. Used to scary swims in the Canyon, I told myself to stay calm and breathe between waves, but I felt panic creeping in. The river at Crystal is about fifty degrees cold, and it was numbing me. Washing through the worst of the waves and aiming desperately toward shore, I saw a few narrow eddies race by like a string of lost opportunities. The good news was that I'd been running empty—there was no one to save but myself. The bad news was I was starting to feel like I couldn't do it.

Just then I looked up and saw hope. It was Whale, standing at the stern of his thirty-three-foot pontoon rig, motoring out from shore at a perfect ferry angle to catch me. He'd been waiting in an eddy below the hole, right where I needed him to be. Although floating fast and low, I still had the sense to admire him in all his glory, coming to the rescue in full control of his boat and the situation—a supreme look of concentration on his face, his blue eyes in a focused squint, his blond hair properly tousled and wild. He looked as brave as George Washington on the storm-whipped Delaware, as alert as a predator about to pounce, as unwavering as if he were my best friend.

His crew tossed a safety line that landed squarely in my arms. "Hold on!" someone yelled. With Whale minding the helm, his crew and passengers pulled me from the river. They motored me to shore, catching and righting my raft along the way and setting me in it. None of my gear had even moved in the rigging. The rescue was over that fast—as if nothing had happened.

But Whale and I knew it had happened. He passed a bottle of whiskey to me in my washed-clean boat. To the applause of his crew and his own silent appreciation, I stood in the rowing well and threw back a few slugs. Although I felt waterlogged and cold to the bone, I'd survived it, and Whale was my hero. I raised the bottle and toasted him and high water.

Back in my hometown tributary, I'm separated by time and distance from Crystal Rapids. I've been retired from commercial boating for fifteen years. Whale has been gone more than half a decade. Near the close of the 1995 river season, with fall in the air and the cold nights upon him, he walked into the ponderosa pine forest outside Flagstaff. He stood alone among the sweet smells of tall trees with desert wind in their needles. As I imagine it, he gazed up at a sight he knew he'd never see again—blue sky above a spiral of conifer branches, wisps of white cloud passing even higher overhead. Then he ended his life with a pistol, and he left no note.

My friends called to ask how well I'd known him. Did I know he'd fought in Vietnam? In the war, he'd been a helicopter door gunner and later a crew chief, lasting a year in the job at a time when life expectancy was about three days. Did I know he'd recently broken up with his long-time girlfriend? How well had he handled that end-of-season thing? One friend said she'd seen him a few weeks earlier in line at the Flagstaff post office. He'd looked a mess, as if he'd been "sleeping in a dumpster." He probably hadn't showered in weeks, a habit perhaps carried over from the river. "But I saw him the next day and he looked fine." He'd looked fine to me, too, the last time I saw him, the first time we truly met. He'd been smiling because he had helped me.

The dry creek bed unfurls a path of secrets before me. I pause when I hear an acorn woodpecker frenetically storing nuts for the winter, and I notice that I've stopped near a low bank of gravelly spheres. Each one is tough and a few inches in diameter, a stone-thrower's dream. I break one in two, the halves parting as easily as a cloud. Inside is a center that's damp and muddy, still soft and pliable. Armored mudballs. In spring, before the wet season is truly over, high flows scoop up mud and roll it around in sand and pebbles like cookie dough in sugar. The result—small, soft spheres with tough exteriors.

Finding armored mudballs is a strange comfort. Although this creek runs dry more than half the year, high flows still define its character. The signs may appear inscrutable at first, but if you take the time you can discover their origins. Water-worn boulders studding the shores, the sandy streambed as rippled as an ocean floor, armored mudballs—all are the patient, loving work of moving water. The channel itself, carved below the surface of the floodplain, is proof positive of cycles and seasons.

Had I known how to read Whale's face as well as I've come to understand the features of this dry creek bed, perhaps I could've returned the favor he did me back in Crystal in 1983. Maybe I could've saved him somehow. But I didn't come through for Whale—none of us did. It's much easier to look inside the armor of a mudball than the skin of a man. We didn't know how to convince him that his despair would pass—if indeed it would have—as high water always comes again. Getting him to believe in the return of a better season would have taken more than a pontoon boat, a rescue line, and a boatman's cool skill.

I turn to head down the creek for home. For the first time, I notice that the sunlight looks different today—muted, thick, and angled low. Straightening up to pay attention, I hear a call from farther off and higher up, a mournful cry from the north. In a moment a single bird flies into view, wings beating furiously, neck outstretched and reaching south. It's a lone goose, calling in autumn.

Past the Edge of Land

Most deltas are partly above and partly below water.

—Dictionary of Geological Terms, Third Edition

The night before Seder falls clear and warm in the Sierra foothills. Only the brightest stars outcompete a brilliant half moon. When I tuck my nine-year-old daughter Rose and her little friend Max to sleep in their tent, moonglow sifts through the mesh window onto their contented faces. They've just spent the day running through April-green grass with the children of other former boatmen. They've also befriended a local dog, a sweet Ridgeback mix who now stands watch outside their tent. Setting out a Sierra Club cup full of drinking water seems to be all that's needed to buy the dog's allegiance as sentry. Guarded by the Ridgeback, stuffed with s'mores and roasted marshmallows, the children let their eyes open and close, open and close, as they slip into sleep.

These kids love coming to Seder as much as I do. This is Rose's fifth year attending, Max's third, and about my tenth in the twenty-three-year tradition of annual guide reunions. Seder is held on a many-acred property owned by a handful of ex-boatman families living above the American River. Each year close to a hundred people turn out for the gathering. We pitch our tents on a ridge overlooking the river, throw together raft runs with dozens of people, and hike among the colors and pungent smells of the oak forest and open slopes with fresh swatches of lupine, poppies, and wild irises.

Today we boated the short stretch of river between Coloma and Lotus, a hectic, impromptu trip with troops of kids commandeering the paddles. We adults had to do little more than pump up the rafts, herd everyone into boats, and pass out lifejackets. As we floated, we dug up old jokes to fuel the paddlers' enthusiasm. Rose's favorite: "Hey, duck, you owe me money!" "That's okay, man, just put it on my bill." When the spring sun grew warm enough, we tossed water between rafts in good-natured battles that sprayed but did not drench. The kids came off the water feeling both completely refreshed and bone-tired.

Tomorrow evening we'll feast for Passover, as we do every year. The dinner involves many courses, with traditional salt, mortar, and matzo to symbolize the trials borne by the Jews in Egypt and during their flight through the desert. Not all of us are Jewish, but as part of our reunion practice, we join together to commemorate the exodus of the Hebrew tribes from Egypt in Biblical times. Always there are many speeches and much toasting, as the Passover dinner goes on for hours and we celebrate another year of friendship and good luck. Always we toast the redemption inherent in the triumph of those who suffer hardship.

After Rose and Max drift off easily in their tent, I return to the campfire near the tables where we'll dine tomorrow night. Old-time guides stand in the firelight, some playing guitar, mandolin, or harmonica. Paper lanterns swing overhead on a rope strung between two oak trees. I sit on a wooden bench between two friends I've known since we began guiding together twenty-five years ago. Michael, on my left, has spent the past few decades working in Oregon, driftboat fishing, and raising two sons. Tim, on my right, has been a professional guide on northern rivers, a crab fisherman off the north coast, and a homebuilder. He and his wife have an eight-year-old son adopted years ago in Honduras.

We've all gathered in response to an invitation mailed to us by the few brave souls who organize Seder. This year they sent a postcard with a photograph of a young 1960s couple standing before a decorated Volkswagen bus. Home-sewn curtains hang in the windows. Painted birds, puffy clouds, and a heart-shaped *LOVE* adorn the sliding door and wheel wells. A few crossbars topping the bus pass as a luggage rack, with baggage tied above. Wearing bell-bottoms, bead necklaces, and bandanna headbands, the young couple hold their fingers in Vs. Not for victory but for peace. She has navel-length blonde hair. His hair curls up in an afro big enough to shame Jimi Hendrix.

How familiar they look. Put lifejackets on those two, and they're us many years ago. Park this van at some river crossing or in the commercial rafting outfitter's lot, and you've established our summer base. Or pitch a tent or tipi, build a treehouse, or set up camp in the back of a pickup truck. There you'd find us, transient laborers gathered for summer guiding work on wild rivers. *The Grapes of Wrath* without the hunger and desperation.

These days we arrive at the American in family vans, station wagons, sport utility vehicles, and pickups fitted with custom camper shells. The carriers on our cars are top-name brands affixed with designer attachments and gear boxes. We carry plastic, own homes, run businesses, and work for corporations. Our children have taken to wearing bell bottoms, but now the pants are called *flares*. Still, with all our changes and the years separating us, we must hold something in common, because we always make it to reunion. It's a tradition that's sure to die hard.

Part of the Sacramento-San Joaquin river system, the American River bears fresh water to the San Francisco Bay estuary. On its way to salt water, the river casts off its load. Cobbles drop out, wherever they're discarded; sand falls at the edges of point bars. Down from its higher slopes, the current slows and loses the last of its burden in sluggish water. Only very fine sediment remains suspended in the river for the length of its journey, outlasting the coarse fraction as hummingbirds outdistance geese in migration. When the silt and mud finally fall, they fill streambeds, push the river from plugged channels, and spread unconfined over the floodplain to spill into marshes, swamps, mudflats, and shallow lakes.

In many rivers, the finest of the fine sediment rides past the edge of land, never settling until it reaches the ocean. There it sinks through salt water, contributing layers of earth to the shallow shelves of the continental margin, building a wedge of earth out to sea. In the rivers of the Sacramento-San Joaquin, the fine sediment brought down from the American, Stanislaus, Tuolumne, Mokelumne, Consumnes—rivers draining the Sierra Nevada—sinks through the dark, cold receiving water of the estuary, constructing a delta many miles in from the ocean. In this inland realm, where sediment drifts into place and fresh water tumbles together with salt water, the right mix of temperature and chemicals grows tiny aqueous plants that foster a great, green world. Minute animals thrive, nourishing bigger beasts up the food chain—herring, smelt, bass, salmon.

Whether out at sea or in the estuary, deltas take their name from the Greek letter Δ, a triangle whose apex points upstream like an arrow and whose base spreads out with the fanning river. In practice, though, deltas are varied and diverse in form. Some are called *birdsfoot* for their splayed geometry with multiple talons, or *lobate* for their deeply divided channels. Some are *arcuate*, or fan shaped. The language of deltas goes on and on. Where people have settled the delta, a special vocabulary evolves to name its many parts: *bypass, island, slough, cutoff, canal, channel, levee, dike, sink, tract, wasteway, flat, pond, marsh*. A journey through the Sacramento-San Joaquin delta crosses levees, spans bridges, passes marinas full of power boats, and enters towns so old and tenacious they seem to have grown up with the land. It's not one world—it's thousands, so filleted that it appears the river carved the earth into pieces rather than building it up from scratch.

People have always settled these fertile water-land interfaces as fast as they can accumulate, with farming tools and crop seed, fishing boats and hook. And yet we don't seem to know the delta on sight, the way we recognize the beach or the mountains. All are the products of geologic and geomorphic processes, but the delta seems the most modest about it. Its features sprawl somewhat endlessly, as hard to follow as the many legs of a cross-country race. One can live blithely on top of delta sediments without having a clue about it. When asked once by a friend the way to Antioch, a Sacramento River town, I began my reply by waving east and saying, "It's over in the delta." No more than fifty miles from our home valley. My friend looked both annoyed and puzzled. "Where's the delta?" she asked.

On the morning of Seder, I walk over the hill to visit my friend Jimbo, who lives in a trailer on the property above the American River. While he was away last August, a fire begun by a lightning strike claimed the home he'd been building. Jimbo, a veteran whitewater guide, had been away since April working on rivers in Arizona and Idaho. He owns the land that encompasses his homesite, but he rarely stays for long. He's more likely to be found rowing to hot springs on the Middle Fork of the Salmon River or leading a hike in Grand Canyon than eating breakfast in his little trailer above the river.

Storms have always come to these hills, but the one that torched Jimbo's house hit after midnight, with concentrated fury. Great flashes of light filled the sky, bright enough for some celestial photo shoot. Wind shook the homes perched above the river. When thunder clapped at their windows, one family leapt screaming from deep sleep, even though they've lived on their ridge for many years and are not nervous types. One friend who lives across the river canyon happened to see the lightning touch earth at three in the morning. Jimbo is thoroughly unsurprised by this. "Lots of people were up with the storm that night." And lots of people here are friends, former guides who never fully left the river.

From the three yellow pines where the lightning hit, flames streamed about a hundred yards to Jimbo's modest home. Fire crews arrived at the remote hillside property within the half hour and, even in a landscape parched by seven years of drought, managed to isolate the blaze to five or so acres centered on the struck pines. Neighbors mobilized, too, with hoes, rakes, even heavy equipment, taking a stand down the slope from the fire. But despite such exemplary efforts, each wall and board of Jimbo's house burned to utter completion.

Jimbo leads me on an impromptu tour of his charred property. He shows me the bald, seared seams that run crown to ground on the three tall pines. "The trees are still standing," he says. "But imagine what they must have looked like." Bark flying. Intense light. Fused sparks. In spite of the fury that struck them, though, the pines bear only benign scars. They don't seem to have been seared by bolts of fire that traveled the distance from heaven to ground. Instead the trees look as if a master peeler took a thin, hot slicer to the length of the bark, which appears curled and shrunken from a surgical cut. Nice, precise work—hardly the tree-splitting wedges I'd pictured from all accounts.

At the homesite, we kick at cold coals and survey the twisted remnants of a house: steel supports, homemade concrete footings, two burned-out refrigerators, window frames. Up the hill a hundred yards, Jimbo's key piece of river equipment, a yellow kayak, leans in the fork of a madrone near his small travel trailer. He changed his plans to store the kayak under his house last summer and instead took the boat to Idaho with him. Now he says, "Luckily the fire didn't get my kayak."

Alternately chuckling and lamenting, Jimbo digs through the burned rubble, displaying each ruined item and telling its history. "This is the chainsaw my brother gave me. Look how the heat warped the blade!" He tosses it back to the ground. "We never could get it started." With his toe, he digs through a cardboard box of linoleum. "I'd planned to finish a floor with these tiles. Now I don't think I will—they look terrible." Picking up two deformed metal poles, he cries, "My ski poles!" He drops them and shrugs.

I ask, "Will you rebuild?"

He scans the changed landscape. "I'm considering it," he says. "But maybe I'm lucky I lost it. I kind of like it better now—it's simpler."

A few hours before the feast, I join a group of friends hiking down the hill to the river. Elizabeth, a former California and Oregon boater, now a watercolor artist settled in Colorado. Sue B, a long-time Grand Canyon guide turned audiologist, who built her own home in Arizona. Gail, an ex-Idaho guide turned family counselor, who's driven from Oregon with her newly adopted infant, Isabel. Now Isabel wears a sunhat and rides in a pack on her mother's back. I encourage Rose and Max to join us for the hike, but they find a narrow trail to run along with some other kids. "We'll see you back up top!"

Last year's tall grass lies gray and slick over a carpet of green seedlings. Our staying vertical on the matted plants requires some effort, and we alternately step and slip our way down the slope. By the time we reach the river, we're feeling warmed by our hiking and ready to dip our feet into an eddy. Elizabeth and Sue B follow Gail to a tiny spit of sand where they'll walk Isabel and test a wading pool. Sensing that I may need total immersion, I try a deep eddy sheltered by rock ledges. First my feet into the river, then my ankles, then my pants cuffed and the water up to my calves. Soon I'm out of my clothes and ducking into the spring-cold pool. Under water for no more than a second, I hop out hooting and shivering and dress quickly, my skin tingling under fabric. My nerve endings haven't felt so alive since my last dip in a spring-cold river.

I meet my friends on the sandbar, where Elizabeth holds Isabel by her two hands to balance her on the sand, and Sue B and Gail toss pebbles into the river for the girl's entertainment. We linger in the spring sun, not talking much, then slowly prepare to climb back up the hill. As we begin our ascent, we hear laughter rise behind us. Four rafts spin into view. The boaters, decked out in wetsuits and synthetics, dip their paddles idly while talking among themselves. Happy chatter fills the air above the water.

We stand stock still, all of us transported by the sight of rafts hovering in current.

Then Gail turns so Isabel can see from her backpack. "Look, Isa. Boats!"

Isabel points with her tiny finger. "Doats!"

One by one, the rafts draw downstream into a set of small standing waves. Each boater sits alert and poised. They accelerate as they move from pooled water into whitewater. The boats, carrying no significant loads, dance over the waves. Dip and rise. Dip and rise. We watch until they round a downstream bend and the river flows wild and private again.

We make our way toward the crest of the hill and pause at a wooden cross standing in freshly dug earth. A carved heart marks the junction of the two pieces of unweathered wood that form the cross. Nearby lies a bouquet of garden flowers. This offering memorializes a young boatman couple killed last summer in a head-on automobile collision near Fallon, Nevada. Returning from a river trip in Utah, the couple hit a drunk driver who'd crossed the centerline of the highway into their lane. Friends riding in the back seat survived but sustained serious injuries. One broke his neck and now suffers brain damage; the other's extensive fractures are gradually being repaired through the surgical insertion of numerous metal plates.

Families and friends of the deceased have found the grief at times unbearable. "Carrying on has been hard," said an aunt, one of the retired boatmen who live on this hill. "They were giving in so many ways." The couple had bought land nearby, where they'd planned to build a home.

I stand at the cross with Elizabeth, Gail, Isa, and Sue B, staring at the broad sweep of rapids below. We can hear the rush of water when the wind blows just right.

After a while, I say, "It's sad."

Gail nods. "And scary. Think how many times we've all made that same trip through Nevada. Or between Idaho and the Grand."

"True. And in the middle of the night."

We're tempted to remain and reflect, but we can't do it forever. After all, we have a feast to attend. Feeling both robbed and blessed, we continue up the hill. We follow each other single file on the hand-built trail leading back to the top.

In scientific notation, we use the Greek letter Δ to denote changes in value. To write Δx in an equation is to record a change in the number x. It's a wonderful shorthand that works not only for pure math but also for physical phenomena such as rivers. A stream reaching the end of its journey fans out into the numerous distributaries of the delta. If one channel splits into three, or nine, or fourteen, then we call it a Δ *channel* of two, or eight, or thirteen. Or if we have too many channels to count, the Δ value will be higher than we can figure, a quantity whose size we have to accept on faith. At the seaward edge of the delta, the river leaves its channels altogether and plunges into the ocean. Then the number of channels plummets to zero, giving us the biggest Δ of all.

At Seder, our presiding *rabbi*—really former Idaho boatman Barry, who leads the ritual—reigns over the proceedings. He sits before us in a folding chair on a plywood platform. As we begin our meal, he asks us to share any values we learned from the river. He starts the sharing: "I'd like to start with the value of hard work. Before we were guides, none of us ever rowed fully loaded rafts on miles of flat water with our backs to a strong upstream wind." Applause and cheers. "We didn't grow up carrying boxes and bags to and from camp every morning like mules." Hoots and "Go, Barry!" "We hadn't worked with ropes to tie loads like packers. Or prepared three meals a day for a crowd like logging camp cooks." From the river, with its natural challenges of weather, wind, and distance, we learned to work hard.

"We learned to play hard, too," says Barry, acknowledging all the kayaking, hiking, exploring, and carousing we did on our days off. "Here's to working hard and playing hard." We raise our glasses of the northern California wine contributed by Snake, former Utah boatman turned winemaker. Snake brought several cases of his favorite reds and whites, which he's distributed to every table. Before the feast is over, all the bottles will be drained.

Next Jann, a business consultant who used to guide in Idaho, introduces the value of tolerance. We toast the years spent thrown together with the people in that 18-foot ring of rubber known as a raft. Then Jann adds, "And I have to add dietary tolerance. I've been on river trips where we had nothing to eat but fruit. I've been on trips that offered only blueberry granola for breakfast. I planned a trip myself where everyone had to eat canned mackerel every day for a week." She lifts her glass. "I hope everyone has forgiven me as I forgive them."

We cry, "We forgive you!" and the torch passes. Craig toasts the value of river time. "Where days are not measured by the second hand of a watch but by the natural rhythm of the earth." Elizabeth stands and reminds us of the dam at New Melones, which took the beloved Stanislaus River from us so many years ago. "It was tough," says Elizabeth. "But we learned to go forward after a huge loss." We nod our heads and share a silent moment.

Then Michael toasts the value of community, both among our river friends and in the lives we've made since we retired from guiding. "We were told there was life after the river. We figured it wasn't much of a life." Roars of approval. "But here we are, with our families and old friends, and I've got to say it's great." With that we finish our toasting and dig into the mounds of barbecued chicken, rice, and vegetable salad on our plates.

Soon the children, who've been sneaking bites all through the toasting, finish their meals and run off to play on a couple of rope swings hanging from the oak trees. As the adults feast, I think of another value, actually more of a skill, that I acquired while guiding: learning to say good-bye. Every week or so, we greeted people who'd come from all over the country, even the world, to travel downriver with us. We shared places we venerated with these strangers—within days they felt like intimates. Then in four days, or five, or a week or two weeks, the people returned home, and we guides shuttled back to put-in to meet another group of strangers. We said a lot of farewells, and I for one figured I'd gotten pretty good at it.

We continue feasting until after dark, then torch the firewood that's been piled since midafternoon in a rock ring. Gathering around the fire, we collect in small groups for conversation, and some pull out instruments to play folk and bluegrass. The younger kids find downed branches for roasting marshmallows, and a handful of teenagers experiment with melting plastic cutlery at the edges of the flames. Soon enough, they've finished with that and begun an "all-night dance party" that endures for forty-five minutes.

The moon rises, lambent and white, over the tops of oaks. Although I don't want the evening to end, I feel the night's cold sneak into my bones like an illness. Families and couples are drifting off toward their camps. When Max and Rose return to me, too exhausted to speak, I lead them and the Ridgeback to our tents for the night.

The morning after Seder, I awaken to a ghostly shroud of fog that's stolen inland overnight. A chill has settled on the hill above the American River. Today I'll pack up my camp, dust off the kids, and return home. My friends from out of state will catch rides to various airports and flights to faraway cities and towns. Back to our lives off the river. For Rose, Max, and me, going home means driving toward the ocean, past the levees and wasteways of the delta, out toward the Pacific, and to our valley.

Rose and Max wake feeling adamant. They don't want to leave this place. They'd rather pal around with their little Ridgeback and, when afternoon comes, go on another goofy rafting trip where the adults tell duck jokes and start water fights. The kids lie around in their bags, putting off seeing their friends packing up. We may not be ready to roll up our tents and stuff our sleeping bags, either, but eventually we do it.

The time comes to go, after many hugs all around and promises to return next year. As we drive away and cruise down the hill, I remind myself that our leave-taking is only temporary. The good-byes may pile like little griefs onto the big losses, but most of us have been really lucky. We'll meet again, if not next year, then some time in the years to come. If not here at the American, then on some other river or beyond. We survived huge change years ago—leaving the river for the first time and redefining our day-to-day lives. And we've proved that good-bye, farewell, and see you later are only temporary. We may no longer live in Volkswagen buses with curtains in the windows, but the word LOVE could still be emblazoned on our car doors.

We reach the bottom of the hill, cross the bridge over the river, and bend like the current downstream.

Acknowledgments

Even a book as slender as this one requires the work of many hearts and minds—so much so, I'm astounded by the humanity represented in a single shelf of books. For having gained that appreciation, I'm grateful for having undertaken this endeavor.

There isn't room in these pages to thank everyone who supported me through the process of bringing these essays to print. I owe a huge debt to my friends, my community, and each member of my family, without whom this business of writing would be far lonelier. Special thanks to my gifted writer friends Susan Bono, Arthur Dawson, Lin Marie DeVincent, Connie Gray, Terry McNeely, Louise Teal, and Kathryn Wilder. Your careful reading of some early drafts helped immensely, and I took all of your advice.

Thanks also to:

My agents, Michael Larsen and Elizabeth Pomada, two especially fine people who helped me shape an idea into a proposal.

My editor and publisher Kathleen Hughes, who saw the merit in this book as it was still taking form.

Elizabeth Black, insightful and accomplished watercolor artist, who provided the cover painting.

Reva Solomon, whose brilliant coaching work has inspired me.

My exceptional, courageous, faithful river friends—your story never ends.